Biomedical Waste Management in the COVID-19 Era

Waste and the Environment: Underlying Burdens and Management Strategies

Biomedical Waste Management in the COVID-19 Era

A Global Scenario

Edited by

Vivek Kumar Morya
Senior R&D Scientist, Hallym University, Chuncheon-si, Gangwon-do, Republic of Korea

Raj Morya
Postdoctoral Researcher, Laboratory of Environmental Biotechnology (LBE), National Research Institute for Agriculture, Food and Environment (INRAE), Narbonne, France

Abhay Raj
Principal Scientist, CSIR-Indian Institute of Toxicology Research, Lucknow, Uttar Pradesh, India

Sunil Kumar
Head, Centre for Industrial Ecology and Circular Economy, CSIR-National Environmental Engineering Research Institute, Nagpur, Maharashtra, India; Adjunct Professor, University of Tasmania, Hobart, TAS, Australia

ELSEVIER

Elsevier
Radarweg 29, PO Box 211, 1000 AE Amsterdam, Netherlands
50 Hampshire Street, 5th Floor, Cambridge, MA 02139, United States

ISBN: 978-0-443-21841-5

For Information on all Elsevier publications visit our website at https://www.elsevier.com/books-and-journals

Publisher: Peter Llewellyn
Acquisitions Editor: Pamela Sadhukhan
Editorial Project Manager: Harshita Manchanda
Production Project Manager: Fahmida Sultana
Cover Designer: Raman Kumar

Typeset by Aptara, New Delhi, India

Dedication

This book is dedicated to all frontline healthcare workers, sanitation personnel, and environmental professionals across the world who worked tirelessly during the COVID-19 pandemic. Their courage, resilience, and unwavering commitment to protecting public health and ensuring environmental safety remain a profound source of inspiration. We also dedicate this book to those who lost their lives during the pandemic and to the families who endured the immeasurable loss of their loved ones. Their memories remind us of the human cost of this global crisis and strengthen our resolve to contribute meaningfully toward a safer future. We extend this dedication to our families, whose constant support, patience, and understanding made this journey possible. Their encouragement sustained us throughout this endeavor.

Finally, we dedicate this work to the global scientific and research community, whose relentless pursuit of knowledge continues to drive innovation and develop solutions for a safer, healthier, and more sustainable world.

Contents

Contributors

Muhammad Arshad
Department of Chemical Engineering, College of Engineering, King Khalid University, Abha, Asir, Saudi Arabia

Goo Bo-Jeong
Department of Anatomy and Cell Biology, Lee Gill Ya Cancer and Diabetes Institute, Gachon University, Incheon, South Korea

Vinoth Gnana Chellaiyan Devanbu
Department of Community Medicine, Chettinad Academy of Research and Education, Chettinad Hospital and Research Institute, Kelambakkam, Tamil Nadu, India

Herman Geyer
Department of Geography and Environmental Studies, Stellenbosch University, Stellenbosch, Western Cape, South Africa

Sajjad Hussain
Faculty of Materials and Chemical Engineering, GIK Institute of Engineering Sciences and Technology, Topi, Khyber Pakhtunkhwa, Pakistan

Suman Jagatee
Department of Life Sciences, Rama Devi Women's University, Bhubaneswar, Odisha, India

Basudeba Kar
Centre for Biotechnology, Siksha O Anusandhan (Deemed to be University), Bhubaneswar, Odisha, India

Hammad Khan
Faculty of Materials and Chemical Engineering, GIK Institute of Engineering Sciences and Technology, Topi, Khyber Pakhtunkhwa, Pakistan

Poonam Khanna
School of Public Health, Post Graduate Institute of Medical Education and Research, Chandigarh, Punjab, India

Sunil Kumar
Centre for Industrial Ecology and Circular Economy, CSIR-National Environmental Engineering Research Institute, Nagpur, Maharashtra, India

Mukta Mayee Kumbhar
Department of Life Sciences, Rama Devi Women's University, Bhubaneswar, Odisha, India

Jyotirmayee Lenka
Centre for Biotechnology, Siksha O Anusandhan (Deemed to be University), Bhubaneswar, Odisha, India

Muhammad Usama Majid
Faculty of Materials and Chemical Engineering, GIK Institute of Engineering Sciences and Technology, Topi, Khyber Pakhtunkhwa, Pakistan

Saurabh Mishra
Department of Biochemistry, Chatrapati Shahu Ji Maharaj University, Kanpur, Uttar Pradesh, India

Ramkrishna Mondal
Department of Hospital Administration, All India Institute of Medical Sciences Patna, Patna, Bihar, India

Raj Morya
Postdoctoral Researcher, Laboratory of Environmental Biotechnology (LBE), National Research Institute for Agriculture, Food and Environment (INRAE), Narbonne, France

Vivek Kumar Morya
Department of Orthopedic Surgery, Hallym University Sacred Heart Hospital, Anyang-si, Gyeonggi-do, Republic of Korea

Edet Otto
Department of Environmental Health, Ajayi Crowther University, Oyo, Nigeria

Sujata Priyadarshini
Department of Life Sciences, Rama Devi Women's University, Bhubaneswar, Odisha, India

Alok Raghav
Department of Anatomy and Cell Biology, Lee Gill Ya Cancer and Diabetes Institute, Gachon University, Incheon, South Korea

Abhay Raj
Environmental Microbiology Division, CSIR-Indian Institute of Toxicology Research, Vishvigyan Bhavan 31, Mahatma Gandhi Marg, Lucknow, Uttar Pradesh, India

Harishma Ramesh
Department of Community Medicine, Tagore Medical College and Hospital, Rathinamangalam, Chennai, Tamil Nadu, India

Suprava Sahoo
Centre for Biotechnology, Siksha O Anusandhan (Deemed to be University), Bhubaneswar, Odisha, India

Henry Sawyerr
Department of Environmental Health, Kwara State University, Malete, Nigeria

Manish Singh
Department of Neurosurgery, GSVM Medical College, Kanpur, Uttar Pradesh, India

Ramesha Tariq
Faculty of Materials and Chemical Engineering, GIK Institute of Engineering Sciences and Technology, Topi, Khyber Pakhtunkhwa, Pakistan

Muhammad Usama
Faculty of Materials and Chemical Engineering, GIK Institute of Engineering Sciences and Technology, Topi, Khyber Pakhtunkhwa, Pakistan

Nida Zafar
Faculty of Materials and Chemical Engineering, GIK Institute of Engineering Sciences and Technology, Topi, Khyber Pakhtunkhwa, Pakistan

Editors' Biographies

Dr. Vivek Kumar Morya is a Senior R&D Scientist at Hallym University, Chuncheon, Gangwon, Republic of Korea, where he works at the intersection of regenerative medicine, orthobiologics, and digital health. His research focuses on musculoskeletal regeneration, with particular emphasis on conditions such as tendinopathy and osteosarcoma, alongside the development of advanced cell-free therapies, including exosome-based and tendon-derived stem cell approaches. He is an inventor on several patents and has contributed extensively to the scientific community through numerous peer-reviewed publications in regenerative medicine, biotechnology, and musculoskeletal research. In addition to his academic work, he has also translated innovation into real-world applications, with his microwave-mediated biomedical waste disinfection device successfully commercialized and deployed across multiple regions, including global markets and installations throughout India.

Dr. Raj Morya is a Postdoctoral Researcher at the Laboratory of Environmental Biotechnology (LBE), INRAE, Narbonne, France. His research focuses on waste bioconversion, microbial biotechnology, continuous bioprocess engineering, and sustainable bioproduction systems. He specializes in gas-fed photobioreactors, carbon capture, and process optimization for bioresource recovery.

Dr. Abhay Raj is currently a Principal Scientist at CSIR-IITR, Lucknow, Uttar Pradesh, India. He specializes in environmental microbiology and waste management systems. His work focuses on microbial processes for waste treatment, environmental sustainability, and public health protection. He has been actively involved in interdisciplinary research addressing biomedical and municipal waste management.

Prof. Sunil Kumar is the Head of the Centre for Industrial Ecology and Circular Economy at CSIR–National Environmental Engineering Research Institute (CSIR-NEERI), Nagpur, Maharashtra, India. He is also an Adjunct Professor at the University of Tasmania, Hobart, TAS, Australia, and a Humboldt Senior Fellow. His research focuses on solid and hazardous waste management, municipal solid waste, e-waste, contaminated sites, environmental impact assessment, and wastewater treatment. He has authored more than 300 refereed publications and has contributed significantly to research, policy development, and capacity building in environmental engineering and science.

Editors' Biographies

Preface

The COVID-19 pandemic has left an indelible mark on global healthcare systems, reshaping not only the way patient care is delivered but also how healthcare waste is generated and managed. One of the less visible, yet critically important, consequences of the pandemic has been the dramatic rise in biomedical waste. The widespread and necessary use of personal protective equipment (PPE), diagnostic kits, vaccination materials, and other single-use medical supplies led to an unprecedented increase in both the volume and complexity of healthcare waste streams across the world.

This sudden surge placed immense pressure on existing biomedical waste management systems, many of which were not designed to handle such scale and urgency. In many regions, it exposed gaps in infrastructure, inefficiencies in waste segregation, and limitations in treatment and disposal capacity. At the same time, it brought to light the essential role of effective waste management in safeguarding not just healthcare workers, but also the broader community and the environment. Biomedical waste, particularly its infectious and hazardous components, carries significant risks if not handled properly. During the pandemic, concerns surrounding improper segregation, unsafe handling, delayed transportation, and inadequate disposal became more pronounced. These challenges increased the potential for secondary disease transmission, environmental contamination, and occupational exposure among frontline workers. Such realities underscored the urgent need for more resilient, adaptive, and sustainable approaches to biomedical waste management—approaches that integrate scientific innovation, robust regulatory frameworks, and efficient operational practices.

It is within this context that *Biomedical Waste Management in the COVID-19 Era* has been conceived. This book aims to provide a comprehensive, multidisciplinary perspective on the evolving challenges and opportunities in this field. By bringing together contributions from researchers, academicians, and practitioners, it offers a collective understanding of how biomedical waste management has transformed during one of the most significant global health crises in modern history. The chapters in this volume explore a wide range of critical themes, including changes in waste generation patterns during and after the pandemic, advancements in waste segregation and treatment technologies, environmental and public health implications, and global best practices. The book also examines region-specific challenges and presents forward-looking strategies to enhance sustainability, preparedness, and resilience in the face of future healthcare emergencies. The primary goal of this work is to bridge the gap between traditional waste management practices and the evolving demands posed by large-scale public health crises. It is designed to serve as a valuable resource for a diverse audience, including researchers, healthcare professionals, environmental engineers, policymakers, and industry stakeholders who are working toward safer, more efficient, and sustainable waste management systems.

We would like to express our sincere gratitude to all contributing authors for their dedication, expertise, and collaborative spirit throughout the development of this book. Their contributions have been instrumental in shaping a resource that is both scientifically rigorous and practically relevant. We also extend our heartfelt appreciation to the Elsevier editorial and production team for their continuous support, guidance, and professionalism at every stage of this journey.

We hope that this book will not only advance knowledge in the field of biomedical waste management but also inspire meaningful action. By fostering awareness, encouraging innovation, and supporting informed decision-making, we believe this work can contribute to global efforts aimed at protecting public health and preserving environmental integrity in an increasingly complex and interconnected world.

Acknowledgments

We would like to sincerely thank all the contributors who made this book possible. This work is truly the result of their dedication, hard work, and valuable insights. Each chapter reflects the effort and expertise of the authors, and we are deeply grateful for their commitment.

We especially appreciate all the authors for their cooperation, timely contributions, and patience throughout the writing and review process. Their support and collaboration helped shape this book into a meaningful and comprehensive resource on biomedical waste management during the COVID-19 era. We also extend our gratitude to the Elsevier editorial and production team for their continuous guidance and support. Their professionalism and coordination played a key role in bringing this book to completion. Our thanks also go to our institutions and colleagues for their encouragement along the way. A special and heartfelt thanks to our families for their understanding, patience, and constant support throughout this journey.

Finally, we would like to acknowledge, with deep respect and gratitude, the healthcare workers, sanitation staff, and environmental professionals across the globe. During the COVID-19 pandemic, they stood at the forefront under extraordinarily challenging and often high-risk conditions, ensuring not only patient care but also the safe handling, segregation, and disposal of biomedical waste. Their work, though frequently unseen, was critical in preventing secondary infections, protecting communities, and maintaining environmental safety during a time of global crisis. Their unwavering commitment, resilience, and sense of responsibility highlighted the vital role of effective biomedical waste management systems in public health infrastructure. Witnessing their dedication and the challenges they faced made the significance of this field even more evident to us. Their efforts have not only safeguarded lives but have also inspired the development of this work, reinforcing the need for continued research, innovation, and awareness in biomedical waste management.

Introduction

Biomedical waste management has always been a vital pillar of public health and environmental protection. It ensures that the byproducts of healthcare, often invisible to the public, do not become sources of harm. However, the emergence of the COVID-19 pandemic transformed this already complex domain into a global challenge of unprecedented scale. Almost overnight, the volume, composition, and urgency of biomedical waste management changed dramatically across the world.

The pandemic led to an extraordinary increase in the use of personal protective equipment (PPE), diagnostic kits, vaccination materials, and a wide range of disposable medical supplies. While these tools were essential for saving lives and controlling the spread of infection, they also generated an enormous amount of waste. From large hospitals in developed countries to small healthcare centers in resource-limited settings, the surge in biomedical waste became a shared global concern, cutting across geographic and economic boundaries. This rapid escalation placed immense strain on existing waste management systems. Many infrastructures, particularly in developing regions, were not designed to manage such high volumes of infectious and hazardous waste. Even in well-established systems, gaps became evident—ranging from improper segregation at the source to limitations in treatment capacity and logistical challenges in collection and transportation. These pressures exposed vulnerabilities that had long existed but had never been tested at such a scale.

At its core, biomedical waste management is a carefully coordinated process involving multiple stages: segregation, collection, transportation, treatment, and final disposal. Each step is interconnected, and the effectiveness of the entire system depends on the integrity of each component. The pandemic made it clear that even small lapses, such as incorrect segregation or delays in disposal, can significantly increase the risk of infection transmission, environmental contamination, and occupational hazards for healthcare and sanitation workers. This realization has reinforced the importance of building systems that are not only efficient but also resilient, adaptable, and guided by strong regulatory frameworks. In response to these challenges, *Biomedical Waste Management in the COVID-19 Era* has been developed as a comprehensive and multidisciplinary resource. The book brings together the perspectives of researchers, clinicians, environmental experts, and policymakers to provide both foundational understanding and contemporary insights into this evolving field. It aims to move beyond theoretical discussions by addressing real-world challenges and practical solutions that emerged during the pandemic.

The volume begins by exploring how biomedical waste management practices have evolved in the post-COVID landscape, highlighting shifts in waste generation patterns and operational priorities. It then delves into the direct impact of the pandemic on waste-handling practices, emphasizing the critical importance of proper segregation, the first and most fundamental step in ensuring safe and efficient waste management. Subsequent chapters provide detailed discussions of treatment and disposal technologies, ranging from conventional methods to innovative approaches designed to enhance efficiency and reduce environmental impact. The book also presents a global perspective, drawing on case studies and comparative analyses from different regions to showcase best practices and lessons learned under varying healthcare and regulatory conditions.

Importantly, this work does not overlook the broader implications of biomedical waste. It addresses the environmental and public health risks associated with improper waste management, including long-term ecological damage and potential pathways for disease transmission. Special attention is given to case-specific challenges, such as waste handling in informal settlements, where socioeconomic constraints and limited infrastructure further complicate effective management. Looking ahead, the book explores future directions in biomedical waste management, with a focus on innovation, policy development, and sustainability. It highlights the need for forward-thinking strategies that can prepare healthcare systems for future crises while minimizing environmental impact. These perspectives are particularly relevant in a world where global health emergencies are becoming increasingly frequent and complex.

This book is intended to serve as a meaningful resource for researchers, healthcare professionals, environmental engineers, policymakers, and industry stakeholders. By combining scientific knowledge with practical insights, it seeks to contribute to the development of safer, more efficient, and sustainable biomedical waste management systems, systems capable of protecting both human health and the environment in an ever-changing global landscape.

CHAPTER

1 Advancing biomedical waste management: Innovations, ethics, and future trends in a post-COVID era

Hammad Khan[a], Sajjad Hussain[a], Ramesha Tariq[a], Muhammad Usama[a], Muhammad Arshad[b], Nida Zafar[a] and Muhammad Usama Majid[a]

[a] *Faculty of Materials and Chemical Engineering, GIK Institute of Engineering Sciences and Technology, Topi, Khyber Pakhtunkhwa, Pakistan,* [b] *Department of Chemical Engineering, College of Engineering, King Khalid University, Abha, Asir, Saudi Arabia*

1.1 Introduction

The global landscape has been profoundly reshaped by the advent of COVID-19, leaving its mark on 213 countries as of September 20, 2021. This pandemic has resulted in over 229 million confirmed cases and nearly 4.7 million deaths (Singh et al., 2022). Beyond the immediate consequences, the crisis has unveiled persistent environmental concerns, particularly a surge in solid waste generation and reduced recycling efforts. While lockdowns yielded some positive environmental outcomes like cleaner rivers and clearer skies, solid waste management faces challenges (Sharma et al., 2020). However, this pandemic has introduced face masks and disposable gloves as major components of medical waste, featuring nonbiodegradable materials like polyurethane, polypropylene, and latex. These materials accumulate as microplastics, worsening pollution and greenhouse gas emissions. Changes in waste production, especially medical and hazardous waste, pose contamination risks. Effectual waste management and responsible disposal methods are vital in addressing these issues (Attrah et al., 2022). Consequently, effective waste management practices are crucial to mitigate these environmental consequences.

Biomedical waste (BMW) encompasses a variety of materials originating from healthcare, research, and medical development activities. It includes waste generated by various healthcare facilities, such as hospitals, clinics, and veterinary centers. This waste results from activities associated with the diagnosis, treatment, and immunization of both human and animal subjects. Examples of healthcare waste comprise used syringes, needles, sharp medical instruments, used dressings, blood samples, and body parts. Additionally, it encompasses pharmaceutical products, chemical substances, and radioactive materials that are no longer needed within healthcare settings. BMW can also be generated during the production and testing of biological substances

Biomedical Waste Management in the COVID-19 Era: A Global Scenario.
DOI: https://doi.org/10.1016/B978-0-443-21841-5.00006-0

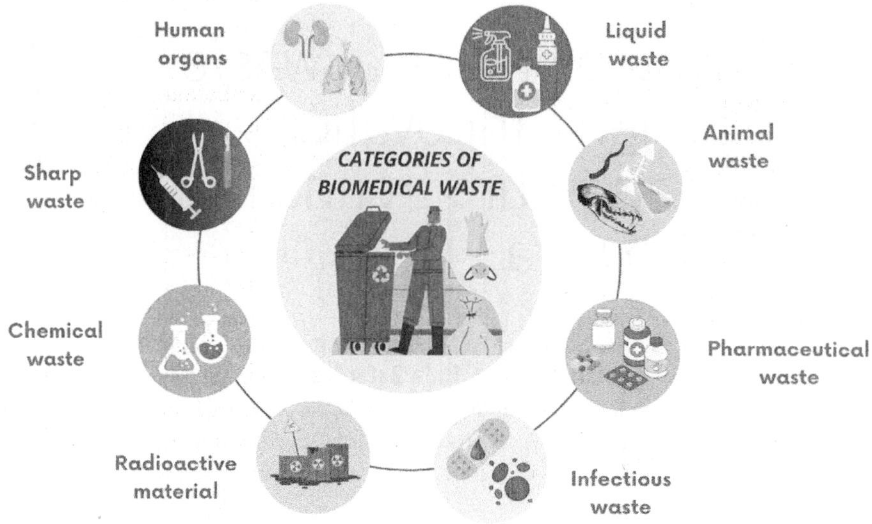

FIGURE 1.1

Categories of biomedical waste.

and in health camps providing medical interventions to communities (Prem Ananth et al., 2010). Furthermore, BMW encompasses pathological waste from the human body, infectious agents, laboratory cultures, and contaminated animal remains from research or healthcare. It also includes chemotherapy waste with cytotoxic agents and discarded medical equipment, excluding noncontaminated single-use supplies (Datta et al., 2018). Fig. 1.1 illustrates the various categories of BMW.

Approximately 10%–25% of medical waste poses a hazardous threat, encompassing infectious, radioactive, toxic, or genotoxic materials. This endangers not only patients and staff but also public health and the environment, making proper waste management essential for risk reduction (Attrah et al., 2022). While most hospital waste is not inherently dangerous, a portion contains infectious microorganisms that can spread diseases in water and air. Therefore, managing hazardous waste is crucial. Developed nations have embraced comprehensive waste practices involving handling, segregation, disinfection, and transportation, all pivotal in healthcare waste management (Thakur & Katoch, 2012). Fig. 1.2 summarizes the current scenarios, options, opportunities, and limitations in dealing with BMW comprehensively.

1.2 Current biomedical waste management landscape

Healthcare waste is a complex mixture that poses challenges for proper management. To simplify this issue, a well-planned management system is crucial. Healthcare waste

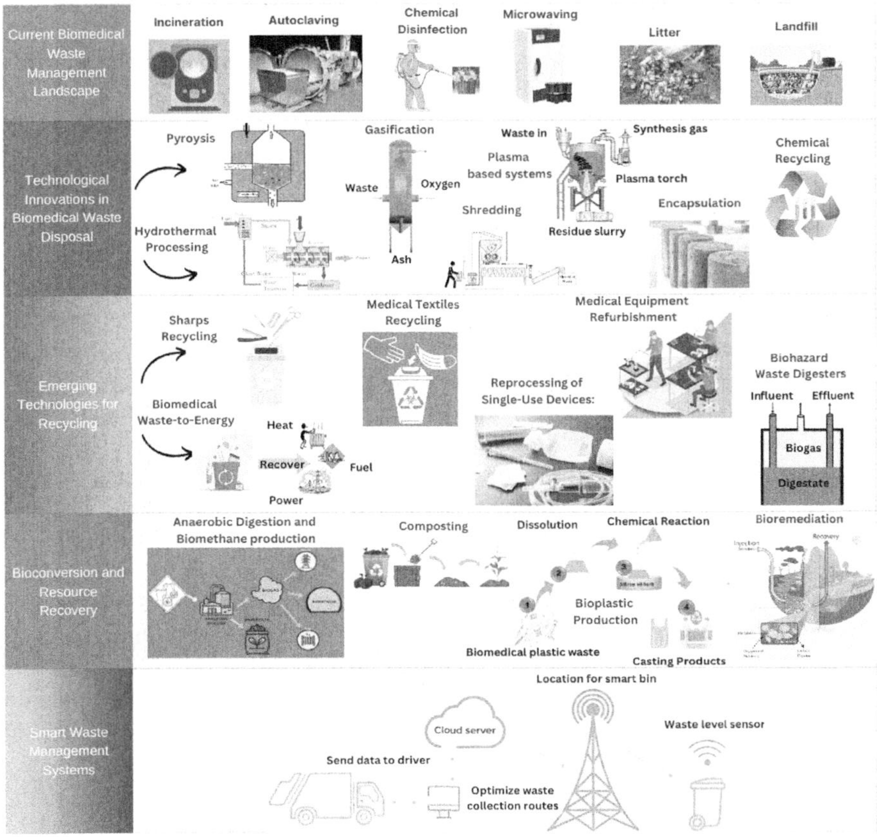

FIGURE 1.2

Current scenarios, options, opportunities, and limitations in dealing with BMW comprehensively. *BMW*, biomedical waste.

management follows a cradle-to-grave approach, involving steps like characterization, quantification, segregation, secure storage, safe transportation, and appropriate treatment. This comprehensive approach ensures the safe disposal of BMW, reducing health and environmental risks. Developed nations strictly regulate waste segregation and disposal, while developing countries struggle with resource constraints and poor sanitation practices, often mixing hazardous waste with general waste, increasing costs and infection risks (Ali et al., 2017). High-income countries generate substantial medical waste, impacting the environment and public health. The manufacturing and disposal of healthcare waste contribute to greenhouse gas emissions and pollution. Various methods are employed for healthcare waste treatment; however, the main technologies are incineration and nonincineration-based methods, each serving specific purposes in healthcare waste management (Ghasemi & Yusuff, 2016).

1.2.1 Incineration

Incineration, a commonly employed waste disposal method, involves the combustion of waste materials at high temperatures to break down and disinfect solid, liquid, and gaseous waste. This process, also known as "thermal treatment," is primarily aimed at managing hazardous waste such as medical waste while harnessing heat energy. A solid waste incinerator is a specialized facility designed for processing waste under specific conditions, resulting in over 80% reduction in solid mass and a 90% decrease in medical waste volume. Operating within the temperature range of 800 to 1450°C (Chu et al., 2023), incineration effectively disintegrates combustible waste components in the presence of excess air. Its advantages include minimal land usage and adaptability to various weather conditions. Incineration, while effective for waste disposal, is controversial due to the emission of harmful pollutants. Facilities emit hazardous substances such as nitrogen oxide, nitrogen dioxide, sulfur dioxide, hydrogen fluoride, hydrogen chloride, gaseous organic substances, carbon monoxide, dust, total organic carbon, heavy metals, and polychlorinated dibenzo-p-dioxins and -furans. These emissions contribute to environmental and health issues, including global warming, acidification, photochemical ozone formation, human toxicity, and ecotoxicity (Dri et al., 2018). Incineration also emits a small amount of CO_2 from fossil carbon and produces potentially harmful byproducts that can impact the ozone layer and local health. It generates solid residues, including bottom ash and air pollution control residues, which contain high levels of inorganic and carbon compounds due to incomplete combustion. Furthermore, constructing, operating, and maintaining incinerators entails substantial costs (Rajan et al., 2019).

1.2.2 Autoclaving

Autoclaving, or steam sterilization, is a widely used method to treat medical waste. This process utilizes high-temperature steam to eradicate microorganisms from waste materials. An autoclave, a high-pressure metal vessel with a sealable door, is at the core of this process. It features pipes and hatches for steam introduction and removal (Ghasemi & Yusuff, 2016; Kollu et al., 2022). These machines resemble pressure cookers and come in three types: Gravity, prevacuum, and retort. Gravity autoclaves remove air through gravity alone, operating at 121°C and 15 psi for 60–90 min. Prevacuum autoclaves use vacuum pumps, reducing the cycle time to 30–60 min at 132°C. Retort-type autoclaves utilize higher steam temperature and pressure. Autoclaves are recommended for certain BMWs, rendering them harmless and unrecognizable for landfills. Hospitals commonly use them to sterilize reusable medical equipment, especially for extremely infectious waste like sharps or microbial cultures. Efficient inactivation of microorganisms in small waste volumes requires a 60 min cycle at a minimum temperature of 121°C and 1 bar (Thakur & Katoch, 2012).

1.2.3 Chemical and microwave disinfection

Chemical treatment, a method involving the introduction of specific substances to medical waste, serves to neutralize or deactivate harmful pathogens. This process primarily results in disinfection, not sterilization. Liquid medical waste, like urine, blood, hospital sewage, or stools, is best treated using chemical disinfection. Commonly used chemicals for this purpose are sodium hypochlorite (5% bleach solution), hydrogen peroxide (30%), diluted active chlorine solution (0.5%), and the Fenton reagent (0.3 g $FeCl_2$ $2H_2O$ in 10 mL 30% H_2O_2). Alternatively, disinfectants like lime, ozone, ammonium salts, and peracetic acid may be employed. Chemical disinfectants offer several advantages, including consistent efficiency, quick action, the ability to use low effective concentrations, and the effective inactivation of bacterial spores and microbe elimination. After treatment, there are both liquid and solid residues to contend with. Liquid residues can be safely disposed of in the sewer system, while solid residues are typically sent to landfills. Nonetheless, chemical methods have limitations, such as potential health risks to personnel from inhaling volatile chemicals, as well as skin and eye irritations (Ojha et al., 2022; Singh et al., 2022).

On the other hand, microwaving is a high-intensity process that utilizes microwaves to heat moisture within waste, working at temperatures ranging from 177 to 540°C, breaking down organic matter in an inert atmosphere (Kollu et al., 2022). Within this method, microbial inactivation results from the thermal effect of electromagnetic radiation within the 300 to 300,000 MHz frequency spectrum. Microwave heating operates through intermolecular heating, with most microorganisms destroyed at approximately 2450 MHz and a wavelength of 12.24 cm (Thakur & Katoch, 2012). The microwave technique offers distinct advantages: Reduced environmental impact with no harmful residue, minimal heat loss, lower energy requirements, and decreased operating temperatures. In strictly regulated processes, specially designed microwave devices have the potential to inactivate COVID-19. However, some limitations exist, such as high initial capital costs, limited applicability to certain waste types, and potential exposure risks (Zimmermann, 2017).

1.2.4 Litter and drainage

Improper handling and disposal of BMW, which includes materials contaminated with infectious or hazardous substances, presents significant environmental and public health risks. While BMW management is often considered a cost-effective liquid waste solution, irresponsible disposal can contaminate water sources and disrupt local ecosystems, emitting foul odors and attracting flies while disrupting beneficial bacterial ecosystems. Cytotoxic waste, in addition to its inherent toxicity, carries mutagenic properties that make it unsuitable for sewage systems (Singh et al., 2001). Illegal dumping of BMW can lead to a range of health problems, including *infectious diseases* resulting from needlestick injuries, *respiratory infections* like streptococcus pneumonia and *tuberculosis, viral infections* such as measles and HIV, and *gastrointestinal infections* like helminths, salmonella, shigella, and cholera, which are transmitted through materials contaminated with infectious feces or vomitus. Viral hepatitis A, B, or C, spread via items contaminated with blood or bodily fluids, as well as *Candida albicans, septicemia*, and bacterial infections resulting from contact with contaminated blood, are also potential outcomes. Inadequate drainage systems can also cause the leaching of toxic substances, endangering aquatic life and downstream communities. Littered BMWs not only mar landscapes but also heighten the risk of exposure to pathogens and dangerous chemicals for both humans and animals. Addressing these critical issues necessitates effective waste management, stringent regulations, and public awareness campaigns (Sivadas et al., 2022).

1.2.5 Landfill

BMW encompasses hazardous materials, infectious substances, and sharps, demanding special handling and disposal to avert contamination and disease spread. Landfilling, the practice of interring this waste in specified locations, has both advantages and drawbacks. A primary benefit of landfilling BMW lies in its ability to isolate potential infectious agents from the broader environment. When adhering to strict regulations, it can reduce the risk of immediate exposure to harmful pathogens, lowering the chances of disease transmission (Chitnis et al., 2005).

Nonetheless, landfilling comes with notable drawbacks. First and foremost, it is an unsustainable choice. BMW in landfills can lead to groundwater contamination, particularly when sites lack protective liners or containment measures (Rajan et al., 2019). Landfilling also harms the environment through fossil resource depletion, extensive land occupation, foul odors, landscape alterations, and diminished amenity value. Furthermore, it contributes to biodiversity displacement and exacerbates global warming, releasing methane (CH_4) gas. Landfilling leads to acidification, eutrophication, and the formation of photochemical ozone primarily from ammonia (NH_3), nitrogen oxides (NO_x), and volatile organic compounds (VOC) leakage. Leachates, byproducts from waste decomposition, can infiltrate the ground, polluting local water sources and posing serious risks to ecosystems and public health. Disease transmission, acute pollution, and accessibility to scavengers and animals are additional concerns (Ghasemi & Yusuff, 2016). Moreover, landfills have finite capacities, necessitating the establishment of new sites, which may face opposition from communities due to safety and environmental concerns. The disposal of sharps and hazardous BMW in landfills further raises safety issues. Stringent safety measures are crucial to minimize these risks (Chitnis et al., 2005).

1.3 Technological innovations in biomedical waste disposal

Medical waste management in many developing nations was more straightforward before the pandemic. However, poor handling worsened COVID-19's spread, endangering healthcare workers, patients, and the environment. Inadequate BMW management risked contaminating groundwater, endangering humans and animals. Proper containment in hospitals, storage facilities, or landfills is crucial to prevent exposure to birds, rodents, stray animals, and people. Neglecting containment, segregation, or incineration results in environmental hazards, releasing harmful particles into the air. Concerns heighten with radioactive particles from diagnostic technologies seeping into landfills, affecting the atmosphere, and causing illnesses (El-Ramady et al., 2021; Ojha et al., 2022).

The World Health Organization (WHO) does not mandate specific medical waste treatment methods, letting countries choose suitable options. Landfilling is discouraged due to environmental concerns. Incineration is the primary method in Korea, with ash residue landfilled. Steam sterilization, chemical treatment, and microwave treatment are also used. In Malaysia, pharmaceutical waste was even found in river water tests, highlighting the need for strict disposal monitoring (Mohamed et al., 2023). Addressing emerging challenges in medical waste management in developing countries is crucial, considering local conditions and limitations. Today, medical waste disposal has advanced, offering cost-effective, environmentally responsible solutions. It has become easier, safer, and more efficient compared to decades ago. In this context, advanced techniques for managing BMW are discussed later.

1.3.1 Pyrolysis

Pyrolysis is a thermal process that breaks down large polymeric molecules in medical waste in an oxygen-limited environment. This method operates at temperatures between 540 and 830°C, which is lower than incineration. Within the chamber, it effectively eliminates harmful substances like dioxins and furans, releasing clean exhaust steam known as syngas. This process is advanced compared to traditional combustion. The clean exhaust steam has multiple applications, including local electricity generation and fermentation (Dharmaraj et al., 2021). The conversion of BMW into valuable products through pyrolysis is illustrated in Fig. 1.3.

Pyrolysis is endothermic, and catalytic pyrolysis uses catalysts to speed up chain breakages. Variations in feedstock and operational conditions result in different products: Solids, liquids, and gases. Pyrolysis can convert medical waste into biofuel, biochar, and liquid fuel, serving as an energy source for generators and contributing to power generation (Fakayode et al., 2020). Pyrolysis is more efficient and eco-friendly than incineration, with lower CO and CO_2 emissions, achieving up to 95% mass reduction. Compact pyrolytic incinerators are commonly used in hospitals due to their controlled air incineration properties, making them suitable for managing various healthcare wastes. However, this technology consumes substantial heat energy and faces challenges in waste rubber treatment development (Jacob et al., 2021).

1.3.2 Gasification

Gasification is a controlled thermal conversion process that takes place in an oxygen-restricted environment, typically at temperatures ranging from 800 to 1600°C. During this transformation, medical and biological waste is turned into synthesis gases,

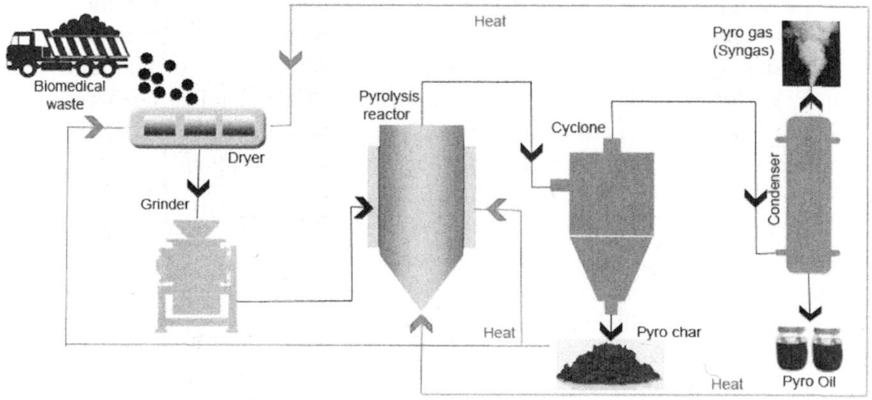

FIGURE 1.3

Pyrolysis of biomedical waste.

primarily consisting of hydrogen (H_2), carbon dioxide (CO_2), carbon monoxide (CO), methane (CH_4), and longer-chain hydrocarbon gases. These resultant gases can be further harnessed for energy production. By subjecting BMW to high temperatures in an oxygen-deprived environment, gasification efficiently and safely eliminates hazardous materials and pathogens. This process not only reduces the environmental impact of waste disposal but also offers a valuable energy source. Gasification of BMW holds great promise in promoting responsible and efficient waste reduction while recovering valuable resources (Zhang et al., 2021).

Gasification processes can be categorized based on the atmosphere used: Air gasification and steam gasification. Air gasification employs an atmosphere containing a mixture of oxygen and nitrogen, resulting in gases with low calorific content. In contrast, steam gasification relies on steam as the medium, producing gases with high hydrogen content through an energy-demanding endothermic process (Chu et al., 2023). A multicriteria analysis demonstrates that gasification is 30% more sustainable than incineration when comparing waste-to-electricity technologies. Syngas obtained through gasification contains higher methane and hydrogen concentrations than pyrolysis but includes impurities. The potential to generate energy, energy carriers, and chemicals from syngas makes gasification an attractive option for revalorizing plastic waste (Mastellone & Zaccariello, 2010). However, plastic waste gasification often generates high-tar gases, some exceeding 160 g/Nm^3, necessitating supplementary gas purification. To mitigate this drawback, gas cleaning systems can be used after gas production or before the usage of gas products. Additionally, materials with sulfur content exceeding 0.73 wt%, such as rubber and polyvinyl chloride (PVC), are unsuitable for gasification (Chu et al., 2023).

1.3.3 Plasma-based systems

Plasma-based systems represent advanced solutions in BMW disposal, tapping into the potential of plasma, the fourth state of matter, to disintegrate waste into smaller particles or even atoms through energy transfer. These systems utilize the intense heat generated by plasma, often exceeding 10,000°F, effectively converting various waste types, including biomedical, municipal, and hazardous waste, into elemental components and valuable byproducts (Gomez et al., 2009). When medical waste encounters the plasma arc, it undergoes pyrolysis, converting into CO, H_2, and hydrocarbons. These gases are incinerated to ensure safe disposal. Plasma-based waste management is distinguished by its high energy levels, densities, and temperatures, facilitating swift material conversion on a smaller scale (Heberlein & Murphy, 2008).

One strategy for managing medical waste is plasma gasification. This method involves breaking down waste using a specialized torch that generates extremely hot plasma without oxygen. This approach improves combustion and reduces the need for additional oxygen, resulting in cleaner, sterile emissions compared to conventional processes. While it effectively manages municipal and hazardous waste, its use for medical waste is still under development, with challenges like significant capital investment, emissions, high energy consumption, and maintenance. Further research is essential to enhance efficiency, maximize electricity generation, and unlock the potential of plasma gasification for medical waste treatment and hydrogen production (Chu et al., 2023; Erdogan & Yilmazoglu, 2021).

Plasma technology plays a pivotal role in valorizing medical waste, producing hydrogen, syngas, and electricity with superior energy efficiency compared to traditional incineration. This process not only significantly reduces waste volume but also eliminates pathogens and hazardous chemicals, rendering the waste noninfectious. However, it is essential to consider the substantial electricity demands, as thermal plasma generators require temperatures as high as 9723°C, and the potential for nitrogen oxide emissions (Erdogan & Yilmazoglu, 2021).

1.3.4 Hydrothermal processing

Hydrothermal treatment, a remarkable process, involves immersing BMW in hot, high-pressure water or other supercritical fluids. In this transformative procedure, organic components in BMW, such as plastics, sewage sludge, and mycelial dregs, undergo hydrolysis, dehydration, condensation, and carbonization. These reactions convert the waste into a range of gaseous, liquid, and solid products, yielding valuable resources like biochar. This method not only reduces waste volume but also offers opportunities for energy recovery and valuable material creation (Shen et al., 2017). Hydrothermal treatment boasts unique qualities that make it invaluable for managing biological waste. It mimics organic solvents' behavior at specific temperatures, aiding in dissolving and catalyzing reactions of nonpolar or weakly polar compounds. Moreover, it accelerates activities requiring acid-base catalysis. Operating between 160 and 300°C, hydrothermal carbonization is ideal for BMW. Notably, this approach excels in generating biochar from BMW, showing immense promise. Consequently, hydrothermal pretreatment has garnered significant attention in managing high-moisture BMW (Czerwińska et al., 2022).

Carbonization transforms waste through thermal decomposition into solid char, gases, and heat, offering numerous benefits. It reduces waste volume, sterilizes hazardous materials, and minimizes harmful emissions compared to traditional incineration. The resulting solid char can serve as an energy source or improve soil. Carbonization is divided into two methods: Dry and wet. Dry carbonization, known as torrefaction, gradually heats waste in an inert environment at 1 atm pressure and 200–300°C. Wet torrefaction, on the other hand, raises the temperature between 180 and 280°C while under 2–10 MPa pressure *(Shen et al., 2017)*. Dry carbonization excels for its simplicity, as it does not require waste pressurization or preheating. Both processes produce heat and high-calorific char, with low chlorine content. However, torrefaction has low calcium content, while wet carbonization yields low ash content. Dry nature medical waste eliminates the need for moisture reduction before torrefaction. Wet carbonization, in contrast, faces challenges like discontinuity, high-pressure reactors, and substantial energy requirements. This process is renowned for energy efficiency, low emissions, controllability, and simplicity. It generates various carbon-based materials, including activated carbon, carbon fiber, carbon spheres, graphite, and carbon nanomaterials. Moreover, the high temperatures involved act as a disinfectant, effectively eradicating viruses and contaminants *(Asim et al., 2021)*.

1.3.5 Sterilization, shredding, and encapsulation

Managing single-use medical items such as syringes, needles, and contaminated materials involves vital practices like *sterilization* and *shredding*. This process ensures the complete removal of microorganisms, pathogens, and contaminants from waste, minimizing potential health risks. Various methods like autoclaving, microwaving, and chemical treatments are employed for sterilization. Autoclaving, which uses high-pressure steam, is a common and effective technique. Microwaving offers a quicker alternative for specific waste types, while chemical sterilization neutralizes pathogens. Proper sterilization safeguards the environment and public health, ensuring responsible BMW management. These procedures reduce contamination risk and aid in the safe disposal and recycling of medical waste. Strict regulations govern the sterilization and disposal of hazardous medical waste in hospitals (Rutala & Weber, 2015; Wang et al., 2020).

Shredding medical waste stands as a critical measure in managing hazardous materials generated by healthcare facilities. This process involves mechanically breaking down medical waste into smaller, more manageable pieces. By doing so, it reduces the risk of accidental exposure and contamination during transportation. This is especially crucial for items like syringes, IV tubing, and sharps, as it enhances safety for staff and waste management personnel. Apart from minimizing waste volume, shredding readies it for further treatment or disposal, making these materials safer for the environment and public health (Ojha et al., 2022). Only properly disinfected waste should undergo the shredding process. Shredders, with shafts and rotating blades, are used in this procedure, reducing waste volume by around 30%. Shredding improves disinfectant interaction by increasing the surface area and reducing confined gaps. It is vital to avoid introducing metallic waste into the shredders to prevent damage to the blades. After shredding, the resulting materials are available for sale to authorized plastic molding agencies (Gautam et al., 2010).

Encapsulation is a process designed to safely contain and isolate hazardous materials following thorough disinfection, thus preventing environmental contamination and safeguarding public health. It is especially preferred for managing sharps, which encompass contaminated materials capable of puncturing conventional waste containers, posing significant risks by spreading harmful elements, such as syringes (Hazardous Waste, 2020; Ojha et al., 2022). The procedure entails crushing or breaking down raw BMW into smaller fragments using a crusher machine. These disinfected fragments are then placed within metallic drums, which are shielded with protective plastic foam, securely sealed, and made puncture-resistant. Subsequently, these encapsulated materials are disposed of in designated landfills. The primary objective is to minimize the risk of pathogenic transmission and chemical exposure during storage and transportation. Effective encapsulation is vital for ensuring the safe handling and disposal of BMW, thereby reducing the potential for infections, injuries, and environmental harm. Encapsulation is widely recognized as a cost-effective, safe, and straightforward approach for healthcare waste disposal, requiring minimal equipment and operational expenses (Jacob et al., 2021). It becomes the method of choice when no other physical or alternative disposal means are available or when the daily generation of BMW surpasses treatment facilities' capacity. This method also played a crucial role during the pandemic, as many countries resorted to it for the emergency management of substantial COVID-19-related BMWs.

1.3.6 Chemical recycling

Chemical recycling is a transformative approach in managing BMW, encompassing pharmaceuticals, laboratory chemicals, and hazardous materials. It employs advanced chemical reactions to render such waste nonhazardous and even beneficial for other applications, mitigating environmental harm and minimizing harmful residues. This method efficiently breaks down complex compounds in BMW, neutralizing harmful substances, crucial for reducing pollution risk linked to traditional waste disposal (Thakur et al., 2023).

Disposable *masks*, part of BMW, can find renewed purpose in energy storage applications through chemical recycling. Fuming sulfuric acid treatment for 2 to 6 min alters their properties, making the surface more durable and conducive to energy storage materials. This transformed surface, rich in hydroxyl (-OH) and sulfonic acid

($-SO_3H$) groups, serves as a superior separator in aqueous rechargeable batteries (Lyu et al., 2024).

Furthermore, chemical recycling can recover precious resources from BMW, such as pharmaceutical waste containing trace amounts of valuable metals. This not only curbs environmental impacts but also promotes resource conservation. While chemical recycling holds great potential in BMW management, safety, cost-effectiveness, and eco-friendliness must be priorities. Ongoing research and development in this field offer exciting prospects for more sustainable and responsible solutions to healthcare waste challenges (Kaya, 2016).

1.4 Emerging technologies for recycling

The modern era's rapid technological progress demands an inventive approach to BMW management, enhancing recycling options. These innovative methods, reducing ecological impact and bolstering sustainability, benefit not only the environment but also the healthcare industry's economy. Innovative technologies are now at the forefront of BMW management, providing eco-friendly solutions. These advancements are crucial in shrinking healthcare facilities' ecological footprint and lessening medical waste's environmental impact. Notably, various groundbreaking techniques have arisen, each specializing in recycling specific BMW categories, advancing the cause of a sustainable healthcare industry (Kaur, 2023).

1.4.1 Pharmaceutical waste recycling

Pharmaceutical waste recycling plays a crucial role in sustainable healthcare and environmental management. This practice encompasses the responsible collection, treatment, and reprocessing of expired or unused medications, curbing the environmental and public health repercussions of improper disposal. Advancements in recycling technologies, especially within healthcare facilities, seek to reduce the environmental footprint by preventing incorrect disposal practices, such as flushing medications or discarding them in regular bins, which can lead to water source contamination, affecting aquatic life, and posing potential risks to humans. Recycling pharmaceuticals also conserves resources, reusing unexpired medications in underserved communities, reducing production needs, and healthcare expenses. Advanced technologies can collect, sort, and sometimes reprocess pharmaceutical waste. Some programs also extend to recycling packaging materials, further minimizing pharmaceutical industry waste (Joseph et al., 2021; Sapkota & Pariatamby, 2023).

Pharmaceutical waste recycling aligns with broader sustainability goals, leading healthcare facilities and pharmaceutical companies to embrace eco-friendly practices. This not only reduces greenhouse gas emissions but also conserves resources, decreasing the need for new medication production. Adopting pharmaceutical waste recycling is vital for a more eco-conscious, sustainable healthcare industry. While significant strides have been made, ongoing research and innovation are necessary to refine methods and expand adoption, ensuring a healthier, environmentally responsible approach to pharmaceutical waste management. BMW recycling encompasses sharps recycling, certain plastics, metals, electronic components from medical devices, and medical textile recycling (Xu et al., 2023).

1.4.1.1 Sharps recycling

The responsible management of BMW includes a crucial aspect—the recycling of medical sharps like needles, syringes, and lancets. This process not only ensures the safe disposal of hazardous medical waste but also promotes environmental sustainability. Sharps recycling begins with the use of specialized containers designed to be puncture-resistant and leak-proof. Once these containers are full, licensed medical waste disposal companies specializing in sharps recycling collect them. These companies employ advanced sterilization and shredding techniques to render the collected sharps nonhazardous, ultimately reducing the volume of BMW sent to landfills or incineration facilities. The benefits of sharps recycling are manifold. It reduces the risk of needlestick injuries, which can transmit bloodborne pathogens to healthcare workers and waste handlers. Additionally, it contributes to environmental preservation by decreasing medical waste in landfills and incinerators, thereby reducing harmful pollutants and greenhouse gas emissions. Sharps recycling epitomizes a sustainable approach to healthcare waste management, ensuring the safety of healthcare professionals and environmental protection. It underscores the increasing awareness of the healthcare industry's need for innovative solutions to minimize ecological impact while maintaining strict safety standards (Blenkharn, 2009).

1.4.1.2 Recycling medical device components: Plastics, metals, and electronics

Although plastics have greatly benefited the medical sector, they pose environmental challenges when mismanaged. Much ends up in marine ecosystems, creating microplastics, while plastic production indirectly contributes to water and air pollution through fossil fuel consumption. Incinerating plastic waste can release harmful chemicals and gases. Unfortunately, a significant portion of nonbiohazard plastic items currently goes unrecycled, causing environmental concerns and a dwindling supply of landfill space. Plastic medical waste recycling plays a pivotal role in the larger landscape of BMW management. It offers a sustainable solution to the escalating problem of healthcare-related plastic waste. The process involves reclaiming and treating plastic waste to create valuable products, with a focus on reclaiming and responsibly reintegrating various plastic materials widely used in the medical field, like IV bags, tubing, syringe casings, and numerous single-use items. The repercussions of plastic medical waste are significant in both environmental and economic terms. The healthcare industry generates substantial amounts of single-use plastics. Poor disposal practices result in pollution, increased landfill use, and wasted resources (Sahoo et al., 2023). Consequently, plastic medical waste recycling addresses multiple pressing issues. Despite efforts to reduce plastic usage, understanding the proper disposal and handling of medical plastics remains crucial. The European Union recognizes the importance of a circular economy and plastic waste reduction, resulting in stringent regulations like landfill bans, extended producer responsibility (EPR), and specific recycling goals.

The recycling journey commences with the identification and sorting of recyclable materials. While manual sorting is an option, automated methods like near-infrared (NIR) and X-ray fluorescence are gaining popularity. NIR is ideal for transparent plastics, while X-ray fluorescence distinguishes PVC from PET based on their organic composition. Various techniques enable plastic separation, including density sorting, surfactant-driven froth flotation, electrostatic sorting, and air-based sorting. Once plastic waste is collected, sorted, and cleaned, it undergoes four common recycling paths: Primary, secondary, tertiary, and quaternary. Recycling plastics involves repurposing discarded plastic to create new products. Four primary recycling processes are primary, secondary, tertiary, and quaternary (Joseph et al., 2021; Lamba et al., 2022).

a) *Primary recycling*: This involves reusing plastic scraps to craft new products mirroring the original material. For example, used PET bottles can become new ones, or recycled plastic can blend with virgin material to maintain quality. Known as a closed-loop system, it is ideal for clean, uncontaminated single-type plastics, as waste seamlessly re-enters production cycles.

b) *Secondary recycling*: Also known as mechanical recycling, this method collects, separates, and cleans plastic waste to eliminate contaminants. However, sustaining mechanical properties can be a challenge due to thermo-mechanical degradation and contamination. Yet, it remains valuable, used in producing flooring tiles and employing various methods like screw extrusion, injection molding, and blow molding.

c) *Tertiary recycling*: Chemical recycling converts plastics into smaller molecules, mainly liquids or gases, for chemical and fuel production. Pyrolysis, involving high-temperature catalyst-driven breakdown, is one approach that yields oils, gases, and hydrochar as products. The oil finds applications like furnace oil, while the gas, known as syngas, can replace natural gas or coal. Chemolysis depolymerizes using chemicals, suitable for PET and polyurethane. Catalysts shorten polymer chains, easing the depolymerization of complex polymers.

d) *Quaternary recycling*: Incineration drastically reduces waste volume and recovers energy, typically for highly contaminated waste unsuitable for regular recycling. An example is tire-derived fuel (TDF). Yet, this method generates waste residues and pollution, often mitigated by various means such as activated carbon or acid neutralization.

Medical waste management offers the chance to recycle metals, particularly in the case of medical instruments and stainless steel (SS) waste. These materials typically contain elements known for their corrosion resistance, like titanium alloys, ceramics, and polymers. SS waste consists mainly of iron-based alloys with chromium (Cr) and nickel (Ni), as well as elements like molybdenum (Mo), manganese, carbon, nitrogen (N), phosphorus, sulfur, and silicon. Cr, Ni, Mo, and N enhance SS's mechanical properties and corrosion resistance. Metals from medical devices play a vital role in healthcare, found in various instruments, prosthetics, and diagnostic tools, enhancing medical treatments with their corrosion resistance, biocompatibility, and strength (Joseph et al., 2021).

Surgical instruments, highly regarded for their precision and reliability in medical procedures, are often crafted from SS. This choice is attributed to its exceptional durability and rust resistance. Recycling SS is economically sound as it retains its properties through remelting, creating a closed-loop system; consequently, SS items should never be haphazardly discarded, emphasizing the need for a methodical approach to segregate and recycle them. Expert instrument technicians analyze waste compositions and categorize SS into two groups: Those for repair and those for recycling (van Straten et al., 2021). When containers are filled, a recycling company collects the SS materials, melts them, and produces new sheet metal. This sheet metal then finds applications in manufacturing instrument components. Any leftover materials are diligently returned to the recycling cycle, minimizing waste generation. Most instruments are disinfected at the supplying hospital site through standardized protocols, except for disposable ones, often exposed to high temperatures of ~90°C.

Titanium, renowned for biocompatibility and lightweight qualities, is a top choice in orthopedic and dental implants. Cobalt-Cr alloys, known for strength and durability, are common in joint replacement implants, providing long-term support. Yet, concerns arise regarding allergies, corrosion, and disposal. Quality control, allergy testing, and responsible recycling are vital (Gautam et al., 2022).

Electronic component recycling within BMW management is vital to mitigate environmental harm and resource waste. Biomedical facilities employ a wide array of electronic equipment, from diagnostic devices to monitoring systems, making the disposal of obsolete or malfunctioning components an environmental and ethical quandary. Recycling these elements minimizes electronic waste and responsibly manages potentially hazardous materials. For example, defibrillator pads, commonly used in emergencies, can be refurbished and reassembled instead of being discarded as electronic waste. Similarly, medical imaging equipment like X-ray machines, when retired, can have their electronic components, like circuit boards and detectors, extracted and refurbished for use in other devices (Shukla et al., 2023). This practice both extends the lifespan of valuable components and reduces environmental impacts linked to electronic waste disposal.

1.4.1.3 Recycling of medical textiles

Recycling medical textiles is a pivotal step toward a sustainable and eco-conscious healthcare industry. These textiles, including surgical gowns, drapes, bed linens, and patient garments, play a crucial role in daily healthcare operations. Unfortunately, their widespread use results in significant waste production. Recycling these textiles offers a comprehensive solution to reduce their environmental impact. Recent years have seen the development of innovative technologies and practices for efficient medical textile recycling. The process involves collecting and segregating used textiles, followed by thorough cleaning, sterilization, and quality control. Once recycled, these textiles can be reintroduced into healthcare settings, reducing the demand for single-use items and conserving resources. This not only eases the financial burden on healthcare facilities but also lowers waste disposal costs (Das et al., 2021; Martínez-Barbosa & Moreno-Corral, 2022).

Furthermore, recycling medical textiles aligns with broader sustainability goals, reducing the strain on landfills and conserving energy and raw materials. It significantly diminishes the ecological footprint of healthcare facilities, promoting environmental responsibility and social accountability. As the healthcare industry recognizes the importance of environmental sustainability, recycling medical textiles emerges as a vital practice, embodying both fiscal prudence and environmental stewardship, demonstrating the potential for positive change in this crucial sector.

1.4.2 Medical equipment refurbishment

Medical equipment refurbishment is aimed at prolonging the lifespan of medical tools while ensuring their safety and effectiveness. This meticulous process involves inspecting, repairing, and, if necessary, upgrading various devices, from diagnostic tools to patient monitors. It brings various benefits, both economic and ecological. Foremost, it helps healthcare facilities make the most of their resources, reducing the need for frequent equipment purchases, which can strain budgets. For instance, a hospital that refurbishes its older MRI machine instead of buying a new one can save high costs, which can then be allocated to other critical areas of patient care. Moreover, medical equipment refurbishment aligns with sustainable healthcare practices. It curtails the disposal of equipment, lessening the environmental footprint. Fewer discarded devices mean reduced manufacturing and less medical waste, contributing to a greener, sustainable healthcare system. An example is the recycling and refurbishing of defibrillators, which reduces electronic waste and conserves resources (Mang et al., 2023; Shukla et al., 2023).

Beyond economic and environmental benefits, refurbishing enhances patient care. It ensures devices work optimally, providing reliable tools for diagnosis and treatment. This promotes patient safety and reduces the risk of diagnostic errors due to equipment malfunctions. For instance, a refurbished ultrasound machine can provide high-quality imaging, aiding accurate diagnosis and treatment.

1.4.3 Biomedical waste-to-energy

Biomedical Waste-to-Energy is an innovative, eco-friendly method for managing BMW and offers an environmentally responsible solution, aiding waste reduction and promoting sustainable healthcare industry energy sources. This approach taps into the energy potential of BMW, which often contains hazardous materials, nonrecyclable plastics, and other nonbiodegradable components. The process involves controlled combustion at high temperatures in specialized facilities. These facilities serve a dual purpose: Reducing waste volume and generating heat or electricity. The energy produced can power the facility itself or contribute to the local energy grid, offering a clean energy source. What makes this approach unique is its ability to address waste management and energy needs simultaneously (Manegdeg et al., 2020).

> Microbial fuel cells (MFCs) are an innovative and sustainable technology poised to revolutionize BMW management. These systems harness microorganisms' metabolic activity to convert organic matter from BMW into electricity. Firstly, MFCs offer eco-friendly waste treatment, utilizing the energy produced during microbial organic compound breakdown. This reduces waste volume, curbing groundwater contamination and disease spread. Secondly, they provide a clean, sustainable power source, ideal for resource-scarce healthcare facilities and regions with unreliable power infrastructure. MFCs present an exciting dual solution to pressing global issues: Proper BMW disposal and renewable energy generation. As the world increasingly prioritizes sustainable waste management and renewable energy, integrating MFCs into healthcare facilities and waste systems shows great promise. However, challenges like scalability, cost-effectiveness, and microbial community optimization must be addressed to unleash MFCs' full potential in transforming BMW into a valuable resource *(Jain et al., 2023).*

Converting waste into valuable energy resources reduces landfill use, incineration, and greenhouse gas emissions, while also effectively treating hazardous BMW. However, it is crucial to adhere to strict environmental standards to prevent harmful emissions during combustion. Careful waste material selection and handling are critical.

1.4.4 Reprocessing of single-use devices

Reprocessing single-use medical devices is driven by a commitment to sustainability and cost-effectiveness. This method involves refurbishing single-use devices like surgical instruments and diagnostic tools after thorough cleaning, sterilization, and quality checks. It extends their lifespan, reducing medical waste and conserving resources. One of the main advantages of reprocessing single-use devices is significant cost savings for healthcare facilities. It eases the financial burden of constant new purchases, making quality healthcare more affordable and accessible. Rigorous regulatory oversight ensures reprocessed devices meet safety and performance standards. Reprocessing also benefits the environment by reducing medical waste, lessening the strain on landfills and incineration facilities (Shukla et al., 2023). It promotes a circular healthcare economy, conserving materials and energy while reducing the industry's carbon footprint. While reprocessing offers solutions to economic and environmental challenges, adherence to strict regulations and quality control is essential for patient safety. Continued research and innovation are necessary to expand reprocessable devices and enhance healthcare sustainability and affordability (Lalman et al., 2023).

1.4.5 Biohazard waste digesters

Biohazard waste digesters offer an eco-friendly solution for managing biohazardous medical waste, utilizing microorganisms and heat to convert it into nonhazardous byproducts. Unlike traditional methods like incineration or landfilling, these digesters provide numerous benefits. Firstly, they reduce environmental impact by eliminating harmful emissions linked to incineration. Through a controlled biological process, these systems help prevent pollution and the release of dangerous substances, safeguarding health and the environment (Zhao et al., 2023). Furthermore, biohazard waste digesters lower waste volume, reducing transportation costs and carbon emissions. They are particularly valuable for healthcare facilities, research labs, and organizations generating biohazardous waste regularly, offering on-site disposal for convenience and efficiency. Additionally, these digesters can produce renewable energy or heat during the digestion process. This not only lessens the environmental impact but also brings potential cost savings and sustainability benefits to institutions using this technology.

1.4.6 Chemical neutralization

Chemical neutralization is vital in safely managing hazardous medical materials. This process employs chemicals to alter waste characteristics, making it safe for disposal. BMW includes used needles, pathological waste, and discarded pharmaceuticals, posing risks to human health and the environment if mishandled. Strong oxidizers, reducing agents, acids, or alkalis are typically used in this process. These chemicals interact with hazardous waste components like microorganisms and toxic substances, rendering them noninfectious and safe. This transformation eliminates threats to public health and the environment. Chemical neutralization is versatile, applicable to various BMW types, and often cost-effective, especially for pharmaceutical or hazardous waste. However, it must be conducted meticulously, ensuring compliance with local disposal regulations. This method plays a vital role in mitigating BMW risks, safeguarding public health, and preventing harmful contaminants from polluting the environment. Proper application is integral to comprehensive BMW management (Kadam et al., 2023).

1.5 Bioconversion and resource recovery

Bioconversion and resource recovery from BMW offer innovative and sustainable solutions for healthcare waste management. BMW, encompassing items like contaminated gloves and expired pharmaceuticals, presents a pressing environmental and public health challenge. Conventional disposal methods, like incineration or landfilling, incur high costs and environmental concerns. Bioconversion employs biological processes, involving microorganisms like bacteria or fungi, to break down organic components in BMW, converting them into valuable resources while minimizing harm (Demirbas, 2011). This approach reduces landfill waste and promotes a circular economy, making healthcare more sustainable and eco-friendly.

1.5.1 Anaerobic digestion and biomethane production

Anaerobic digestion of BMW offers a sustainable and eco-friendly solution that reduces waste volume, harnesses energy potential, and addresses environmental concerns while promoting sustainable energy production and resource recovery. This biological process harnesses the power of microorganisms to break down organic materials in the absence of oxygen, producing biogas and minimizing waste. One application of anaerobic digestion for BMW involves the treatment of discarded pharmaceuticals, which often pose risks to water supplies and ecosystems. By targeting pharmaceuticals, anaerobic digestion systems break down their complex chemical structures into less harmful byproducts. This not only reduces the environmental impact but also ensures that potentially hazardous substances do not contaminate water sources. Another significant application is the digestion of organic components of medical waste, such as surgical waste and biohazard materials. Anaerobic processing breaks down the organic matter into biogas, which can be used for energy generation or heat. This diverts organic waste from landfills and transforms it into a valuable resource (Demirbas, 2011).

> Biomethane production is a promising avenue within anaerobic digestion. Anaerobic digestion generates biogas, primarily composed of methane and carbon dioxide. To turn biogas into biomethane, the process requires biogas upgrading. This step refines the biogas by removing impurities, especially carbon dioxide, resulting in high-purity biomethane *(Adnan et al., 2019)*. This valuable resource can be injected into natural gas pipelines, used as a clean fuel for vehicles, or to power industrial processes and residential heating. Biomethane not only addresses waste management but also contributes to sustainability by reducing reliance on fossil fuels and greenhouse gas emissions, promoting a circular economy.

Additionally, the digestate left after the anaerobic process can be safely used as a soil conditioner or fertilizer, closing the loop in the waste management cycle. In summary, anaerobic digestion of BMW offers a promising solution to mitigate environmental and public health risks associated with improper disposal.

1.5.2 Composting

Composting is a distinctive process that converts organic materials within this waste into nutrient-rich compost, a valuable soil conditioner and organic fertilizer for agriculture. This approach reduces the need for chemical fertilizers, lessening the environmental impact linked with traditional waste disposal methods. BMW encompasses a wide range of items, from used medical supplies to biologically contaminated materials. Unlike the conventional association of composting with food scraps and yard waste, here, it provides an innovative solution for managing a portion of BMW sustainably (Demirbas, 2011). The first step in composting BMW involves meticulously separating nonhazardous components like paper, cardboard, and select plant-based materials from hazardous elements such as needles or biohazardous fluids. The nonhazardous components undergo a controlled composting process, breaking down into nutrient-rich compost suitable for agriculture or horticulture. Success in composting BMW relies on strict adherence to regulatory guidelines and rigorous oversight. Even a minor lapse in waste segregation can lead to contamination risks.

Notably, this approach is not suitable for all BMW types, especially those with infectious or chemically hazardous properties. Composting BMW is a specialized practice, requiring professionals well-versed in waste management and biohazard control to ensure safety and compliance. Environmental benefits of diverting specific BMW from landfills and incineration must be weighed against potential risks, emphasizing the need for cautious implementation and unwavering commitment to rigorous safety protocols (Singh et al., 2001).

1.5.3 Bioremediation

Microorganisms, adept at bioremediation, play a crucial role in rendering BMW safer for disposal and recycling. This is particularly vital when dealing with waste contaminated by pharmaceuticals and hazardous chemicals. Bioremediation employs microorganisms, plants, or enzymes to break down, neutralize, or transform hazardous waste into less harmful substances. Specialized bacteria and fungi excel at metabolizing BMW components, reducing their harmful impact. These microorganisms effectively degrade organic materials, including pathogens, pharmaceutical residues, and contaminants. Plant-based phytoremediation uses metal-accumulating plants as natural filters to concentrate and remove pollutants, often present in discarded medical equipment (Sharma & Dubey, 2022).

Microbial cultures are used to degrade organic materials, where specific microorganism strains break complex compounds into simpler, nonhazardous substances. These microorganisms thrive in organic-rich, moist environments, making them ideal for BMW. This approach not only reduces waste volume but also ensures safe disposal, preventing environmental threats. Enzymatic bioremediation is another example, using specific enzymes like proteases and lipases to break down proteins and lipids in BMW. These enzymes are highly precise, targeting problematic components without affecting nonhazardous materials, reducing disease risk and persistent pollutants. Bioremediation minimizes the environmental footprint by leveraging natural processes, reducing reliance on energy-intensive disposal methods, and supporting a circular economy (Randhawa & Kullar, 2011).

Nonetheless, bioremediation has limitations. It is effective for organic materials but not for nonbiodegradable or inorganic components like plastics and metals. It is time-consuming compared to incineration, and impractical for rapid healthcare waste disposal. It is resource-intensive, requiring specific environmental conditions and resources, adding to costs (Sharma & Dubey, 2022). Safety concerns arise when using live microorganisms; proper containment and precautions are essential to address these concerns.

1.5.4 Bioplastic production

Converting BMW into bioplastics involves transforming healthcare discards, including biohazardous waste and single-use medical equipment, into biodegradable polymers like polylactic acid (PLA) and polyhydroxyalkanoates (PHA). These bioplastics serve as eco-friendly alternatives to petroleum-based plastics across various sectors (Rowan & Laffey, 2021). The process minimizes BMW and lessens the reliance on fossil fuels. By upcycling medical plastics, we reduce healthcare waste's environmental impact and endorse the use of sustainable materials. Bioplastics from BMW benefit packaging, agriculture, and medical device manufacturing, fostering a circular economy and sustainability (Paul et al., 2021). As the world embraces eco-conscious solutions, innovation in bioplastic production from BMW gains momentum. This approach offers promise in tackling waste management issues and promoting sustainable practices in healthcare and plastic industries, shaping a greener future.

1.6 Smart waste management systems

Smart waste management integrates technology to enhance waste processes, making them more efficient and sustainable. Emerging waste management tech leverages the internet of things (IoT) for optimized waste collection. For example, companies use smart bins to monitor garbage levels and schedule collections efficiently. In BMW management, an innovative solution gaining traction is indoor tracking in healthcare facilities. This system utilizes radio-frequency identification (RFID) to monitor and manage waste. RFID enables automatic data recording through wireless communication between transmitters and receivers, offering cost-effectiveness and eco-friendliness. RFID uses color-coded tags to match waste segregation. Components include tags, readers, transceivers, workstations with middleware and a database, supporting multisystem integration. Both hardware and software components allow real-time tracking. Hardware comprises active and passive RFID tags, readers, web, and database servers. Software involves communication drivers, APIs, and a graphical user interface (GUI), while the communication occurs via wireless local area networks (LANs) (Fatimah et al., 2020).

To enhance waste management and boost efficiency, iSmartWMS, a cloud-based IoT sensor system, employs secure end-to-end encryption. Its multitier client-server structure supervises the process via smart trash bins, smart waste truck, iSmartWMS server, database, and landfill station controller (Esmaeilian et al., 2018; Likotiko et al., 2017), as depicted in Fig. 1.4.

The iSmartWMS server, a centralized waste management hub, operates as a remote, dedicated cloud server. It manages communication with smart trash bin

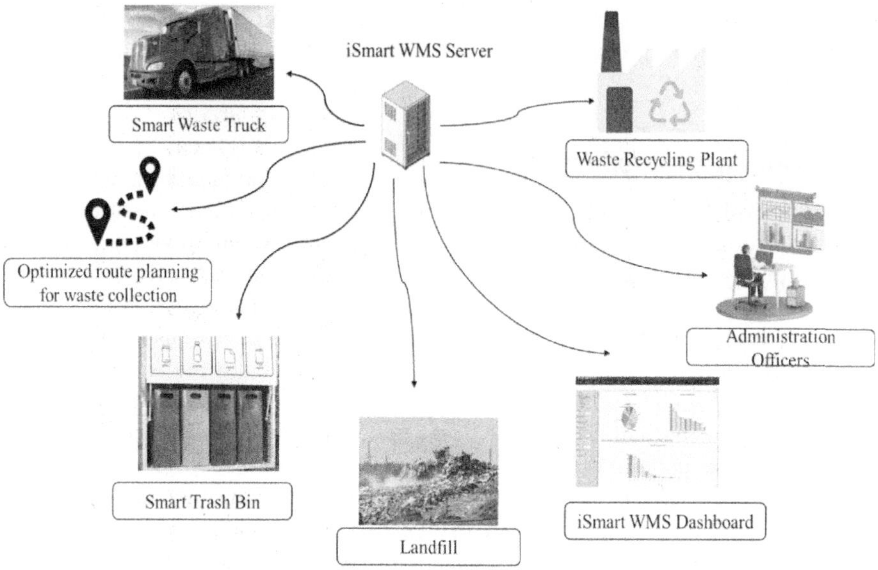

FIGURE 1.4

iSmartWMS server.

controllers and connects to devices like smart waste truck controllers, landfill station controllers, and administration staff through a web interface. The server maintains constant communication, reports failures, runs the smart waste management app, and analyzes waste data. It efficiently stores updates from smart trash bins and controllers in the database, generating waste management alerts, monitoring trash bin fill levels, and instructing smart waste trucks for garbage collection. It also receives real-time data from recycling plant and landfill station controllers, providing a dashboard with information on filled, empty, and under-filled bins. The application server, comprising front-end and back-end components, can be developed using Java or C#.Net technologies and data analytics algorithms using R or MATLAB. The server can be securely hosted on various cloud platforms like CloudX, Cloud Sigma, or other cost-effective options for internet accessibility. iSmartWMS includes a pivotal component: The smart trash bin. Unlike typical bins, it not only gathers waste but also offers real-time data on its fill level and service needs to the world. This bin boasts various sensors, like RFID tags, Ultrasonic sensors, load cells, image sensors, temperature & humidity sensors, and gas sensors. These sensors transmit data regularly to the trash bin controller, which acts as the bin's brain, facilitating communication with the outside world. This consolidated data is wirelessly sent to the iSmartWMS server in a specific format, using wireless repeaters as needed for connectivity.

The smart waste truck (SWT) plays a vital role in smart waste management systems, designed to efficiently gather and transport waste as needed. Unlike traditional garbage collectors, iSmartWMS advocates the use of SWTs for several compelling

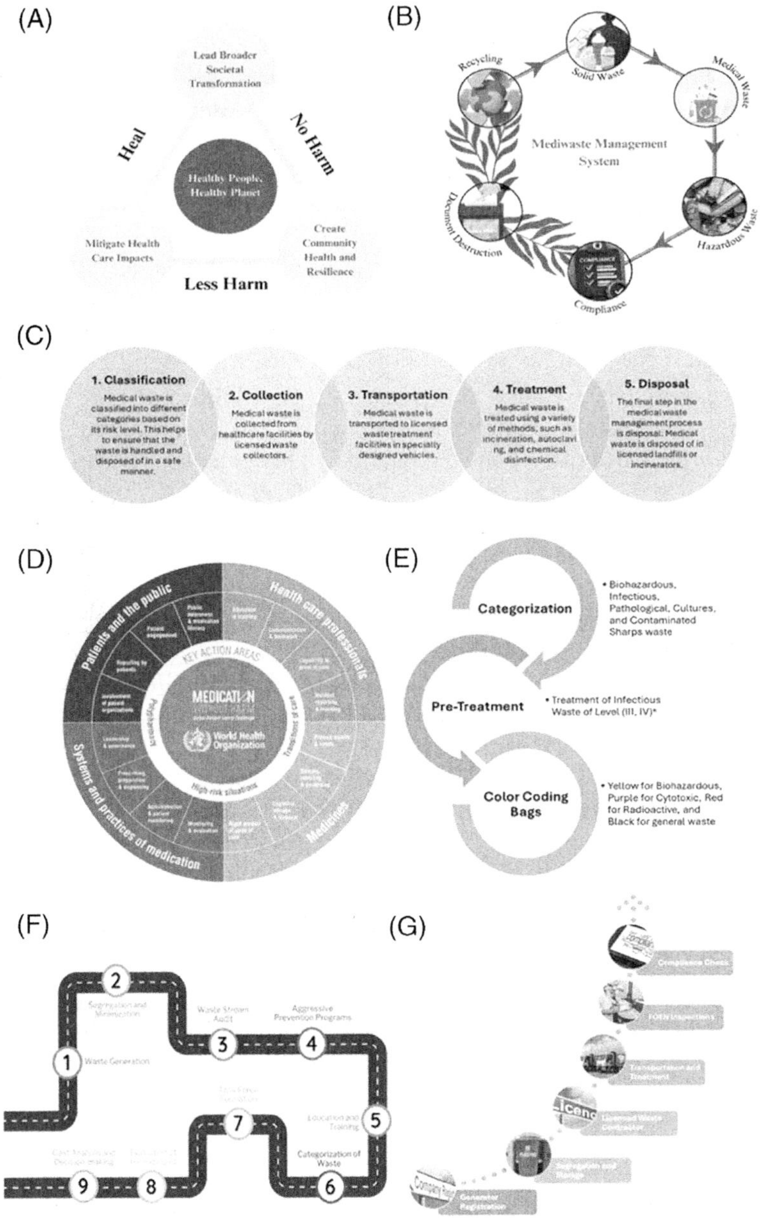

FIGURE 1.5

(A) USA's healthcare without harm (HCWH) system, (B) Mediwaste management system, (C) Avfall Sverige medical waste management, (D) Copyright WHO's medication without harm, (E) Singapore's NEA medical waste management, (F) USA's HCWH process, (G) Switzerland's FOEN medical waste management.

reasons. SWTs feature a robust waste container with a dumpster lift, a robotic arm for automated waste collection, separate compartments for wet and solid waste, a waste level indicator for capacity monitoring, accelerometers, air-quality sensors, infrared cameras, wireless signal monitors, 24/7 secure wireless communication support, GPS, and a client system for real-time communication with the waste management server. These SWTs can securely communicate with the iSmartWMS server, receiving alerts, notifications, and route optimization data while collecting waste from geographically distributed smart trash bins.

The iSmartWMS database, storing vital waste management data, should employ encryption for security. It can be established with open-source databases such as MySQL, MongoDB, PostgreSQL, Redis, or through readily available cloud platforms.

The recycling plant controller's primary responsibility is to provide recycled material information to the iSmartWMS server, which it receives through a weighing scale system or a dedicated operator interface. Daily, the operator inputs required information into the plant controller's user interface, and this data is subsequently transmitted back to the iSmartWMS server as needed. This controller generates a range of critical information related to waste material management, including daily amounts of waste material received and recycled, as well as totals for the past week, month, and year. It also offers data on recycled items, normal operations, alerts, emergency situations, and health status updates.

The landfill controller plays a crucial role in delivering information about landfilled materials to the iSmartWMS server, akin to the recycling plant controller. This data is obtained through a weighing scale system or a dedicated operator interface and is input into the landfill controller user interface system daily, subsequently relayed back to the iSmartWMS server as required. The landfill station controller offers a range of vital information, including daily waste intake, total landfill quantities for the past week, month, and year, as well as details about gas emissions and operational updates.

A proof of concept for the iSmartWMS system was successfully tested in a controlled environment, showcasing robust data analytics, alert generation, and route optimization capabilities (Aithal et al., 2021). Authenticated users can securely access the iSmartWMS server through a user-friendly web interface, leveraging MySQL for the database, ultimately receiving positive feedback from users during lab testing.

1.7 Recycling medical devices

Healthcare e-waste encompasses abandoned electronic medical equipment, alongside conventional healthcare waste such as needles, syringes, and expired medications. Medical facilities employ specialized equipment like sphygmomanometers, electrocardiograms, spectrophotometers, and microscopes, which eventually become discarded. Some of these devices encounter chemicals and biological agents, posing potential health risks. Proper disposal necessitates technical treatment, involving cleaning, recycling, or repairing (Perkins et al., 2014).

Exploration into medical e-waste management is imperative. Medical gadgets constitute a significant portion of the 57.4 million metric tons of global e-waste projected for 2021. Recycling medical equipment holds promise in mitigating electronic waste in healthcare. The evolution of medical device materials, shifting from metals to ceramics and eventually plastics, profoundly impacts the industry (Joseph et al., 2021). Recycling and repurposing medical devices offer a compelling opportunity to extend the life of costly and advanced equipment, such as diagnostic machines, surgical instruments, and patient monitors. Rather than consigning these devices to landfills, healthcare institutions can explore the potential of refurbishment or upgrades, effectively reducing electronic waste while curbing the demand for fresh manufacturing, a process that can deplete resources and harm the environment. These initiatives also resonate with the core principles of the circular economy, focusing on the efficient conservation and reuse of resources.

The practice of recycling and repurposing in the medical field aligns well with the broader goals of resource conservation and e-waste reduction. In this vein, promising initiatives are emerging to introduce innovative recycling technologies and medical equipment refurbishment programs. Nonetheless, medical device recycling does present certain challenges, ranging from data security concerns to device design and regulatory compliance.

- *Single-use medical items:* Advanced recycling technologies have emerged to safely and efficiently reprocess items like syringes and needles. An exemplar of innovation is sharps compliance, which employs microwave sterilization to melt used needles and convert them into new plastic products (Cohoon, 2002).
- *Complex medical devices:* In tandem with the progress in recycling, sophisticated medical equipment, including MRI machines and pacemakers, can now be securely managed. Companies like WEEE recycling solutions have devised methods that disassemble these complex devices into their constituent parts for responsible recycling (Guzzo et al., 2020).
- *Refurbishment:* A sustainable practice involves refurbishing outdated medical equipment and making it available for sale to hospitals and clinics in underserved regions. This not only diminishes the demand for new medical devices but also prolongs the lifespan of existing ones. MedShare is a notable example, actively collecting, restoring, and distributing obsolete medical equipment to facilities in need (Reaching Out 2011).
- *Upcycling:* Creativity finds a place in repurposing old medical equipment into novel goods. For instance, second wind transforms retired surgical tools into unique jewelry and upcycles IV bags into aesthetically pleasing home decor items, contributing to environmental sustainability and ingenuity.

Recycling and repurposing medical devices are a noble endeavor, yet it faces significant hurdles. The foremost obstacle is device design; medical equipment is often created with little thought given to disassembly and refurbishment. Manufacturers must prioritize modularity, upgradability, and standardized interfaces to facilitate efficient and cost-effective recycling and repurposing. Regulatory compliance presents another formidable challenge. Stringent regulations ensure patient safety, making repurposing or recycling seem risky. Clear guidelines from regulatory bodies are crucial to ensure high safety standards during refurbishment and repurposing processes. The willingness of healthcare facilities to engage in recycling and repurposing practices is another hurdle (Kheirabadi & Sheikhi, 2022). Many lack the knowledge and infrastructure needed, coupled with concerns about liability and long-term cost-effectiveness. Overcoming this challenge requires broad educational efforts to inform healthcare professionals about the environmental and financial benefits while providing responsible device handling guidance. Also, medical device recycling must align with FDA regulations to ensure patient safety and legal compliance. Designing gadgets with disassembly and reutilization in mind can reduce electronic waste (Ordway et al., 2020). The industry's eco-friendly design and responsible recycling enhance healthcare's efficiency and sustainability. Addressing these challenges can pave the way for sustainable and environmentally responsible medical device practices.

1.8 Challenges and ethical considerations

The COVID-19 outbreak disrupted medical waste disposal and surgical equipment supply networks. This led to a shift from the traditional "take, make, and dispose" approach to the circular economy (CE) model, reducing the need for new resources. Integrating IoMT, AI, and Big Data in Industry 4.0 aids efficient waste management. Urban areas benefit from smart healthcare waste disposal systems using mobile apps and GPS (Chauhan et al., 2021). Ethical concerns around recycling BMW, especially regarding patient privacy and consent, require thoughtful examination. Recycling BMW, aimed at reducing environmental impact and resource use, raises ethical questions tied to patient privacy and consent. Ethical considerations loom large in BMW recycling, particularly concerning patient privacy and consent (Piasecki et al., 2021).

Privacy is at the core of the primary ethical concern, focusing on the possible infringement of patient confidentiality. BMW often contains sensitive patient information, including personal identifiers and medical histories. Protecting and de-identifying this data is crucial. Patients must grant explicit consent for the recycling of their medical records, diagnostic tests, and devices. An ethical flowchart can guide healthcare organizations in this process. Adherence to ethical guidelines, such as the Health Insurance Portability and Accountability Act (HIPAA) in the United States, is vital for maintaining patient trust and regulatory compliance (Kannan et al., 2024).

Consent is equally vital. Patients trust healthcare providers with their medical data and expect careful handling. Recycling procedures must strictly follow consent protocols. Patients should be informed about the recycling of their medical waste and offered the choice to opt out if they have concerns. Respecting patients' autonomy in handling their BMW upholds ethical standards (Piasecki et al., 2021).

Public perception poses another challenge. BMW recycling can raise public concerns, especially when the processes are not transparent. Healthcare providers and recycling facilities must engage in open communication to build public trust. Demonstrating commitment to ethical standards and patient privacy can alleviate concerns and enhance public trust in BMW recycling (Subramanian et al., 2021).

Solid industrial relationships are essential to tackle ethical challenges in the healthcare waste management sector. A 2023 report from the Healthcare Waste Management Association (HWMA) revealed insufficient cooperation among healthcare providers, waste management firms, and recyclers (Beheshtinia et al., 2023). This deficiency leads to inefficiencies and increased costs. Another report from the World Business Council for Sustainable Development (WBCSD) emphasized the need for increased industry collaboration to establish sustainable BMW recycling procedures. This cooperation can reduce the environmental and public health risks associated with BMW recycling.

Regulatory compliance is another significant hurdle in BMW recycling. Regulatory frameworks must evolve to accommodate healthcare sector recycling nuances while respecting privacy and consent. Inadequate regulations can lead to ethical lapses, highlighted in a 2023 EPA report exposing noncompliance in waste recycling facilities (Newman, 2023). The European Commission's report echoes these concerns, citing inconsistent enforcement of recycling regulations across EU member states. Healthcare facilities must navigate complex local and national laws, while public concerns about safety and privacy hinder BMW recycling. Industry-wide collaboration is crucial to overcoming these hurdles. Regulatory bodies, recycling facilities, and healthcare providers can create a cooperative framework to simplify compliance and boost public trust. This collective effort addresses public concerns, navigates regulatory barriers, and adopts cutting-edge recycling technology while upholding ethical standards. Strong industrial relationships are imperative for success in this endeavor.

1.9 Global case studies

Pharmaceutical mistakes and unsafe pharmaceutical practices are a global health concern, causing significant preventable harm at an estimated annual cost of USD 42 billion. These errors, resulting from human mistakes and system flaws, can have serious or fatal consequences at various stages of the process. The WHO has launched the "*Medication Without Harm*," the third global patient safety challenge, with the aim of reducing the global rate of serious medication-related harm by 50% within the next 5 years (Challenge, W.H.O.G.P.S, 2017). To tackle this urgent issue and enhance pharmaceutical procedure safety, the initiative seeks broad participation from professional associations and WHO member states. The global patient safety challenge encompasses four dimensions outlined in the strategic framework: Patients and the public, healthcare providers, medications and systems, and pharmaceutical practices, each further detailed in four subdomains. Within each domain, the key action areas, including polypharmacy, high-risk situations, and transitions of care, are relevant, forming a central focus.

- The *United States' healthcare without harm (HCWH)* program showcases how cutting-edge BMW disposal and recycling technology can be successfully implemented. HCWH promotes environmentally conscious healthcare practices, emphasizing safe medical waste disposal. Hospitals and healthcare facilities participating in the program have embraced modern waste management techniques like autoclaving and on-site treatment systems, reducing disposal costs and the environmental impact of BMW (Hart, 2022). Health care without harm opposes incineration due to its environmental impact when burning healthcare waste, including solid and regulated medical waste. The process generates toxic air emissions and ash residue, posing risks to both local and distant communities. Incineration is a major source of dioxin and mercury emissions, contradicting environmental sustainability goals. Furthermore, the practice of burning recyclable items in on-site medical waste incinerators goes against ethical considerations for harm prevention in healthcare practices. HCWH emphasizes the need for waste segregation and minimization activities to avoid over-sizing equipment and the associated costs. The program encourages comprehensive audits of various waste streams and products, along with implementing aggressive waste and pollution prevention programs. HCWH underscores the importance of proper waste segregation, identifying waste streams, and educating staff. Additionally, the program raises questions related to organizational/staff issues, facility and operational issues, and on-site treatment technology issues. HCWH places a strong emphasis on considering environmental and fiscal impacts, safety, and repair history when choosing waste treatment technologies. The program also suggests exploring alternatives such as sharps container reuse programs and evaluating off-site treatment facilities with a focus on community and environmental concerns. Cost considerations, including capital equipment costs, labor, utility, and compliance fees, are highlighted by HCWH as integral components of the decision-making process.
- In the pursuit of safer healthcare practices, the *WHO's "Medication without harm"* initiative, along with other commendable projects such as "Mediwaste" and the USA's "HCWH" campaign, offer invaluable lessons and best practices with global relevance. These initiatives underscore the pivotal role of strong leadership and collaboration among regulators, policymakers, and healthcare professionals in driving systemic change. They demonstrate the benefits of modern drug management systems, standardized procedures, and comprehensive training programs to significantly reduce errors and enhance patient safety. Facilitating cross-border experiences and data sharing further promotes the global adoption of effective solutions. These lessons highlight the diverse strategies essential

for improving pharmaceutical safety and reducing harm in healthcare systems universally (Challenge, W.H.O.G.P.S, 2017).

- *Singapore,* a global leader in biomedical recycling, achieved a remarkable milestone in 2021. Over 90% of the nation's BMWs were successfully recycled. The intricacies and potential hazards associated with BMW make this achievement truly outstanding. Several factors contributed to this success, with the unwavering commitment of the government to environmental sustainability playing a crucial role. The BMW (Management and Disposal) regulations, mandating the proper treatment of BMW from segregation to disposal, are significant laws supporting this endeavor. Singapore's BMW recycling program offers financial incentives to healthcare facilities and waste management companies to promote recycling. Additionally, the BMW research and development fund was established to advance cutting-edge recycling technology. Crucially, collaboration among the public, businesses, and academia has been pivotal. Public-private partnerships, like the one between Veolia Environmental Services Singapore and the National Environment Agency (NEA), have led to the creation of state-of-the-art facilities. This system categorizes waste into various types, including biohazardous, infectious, pathological, and contaminated sharps waste. It mandates specific handling and disposal methods, such as treating infectious waste before classifying it as biohazardous. NEA regulations ensure the proper separation of mercury-contaminated waste from general and hazardous industrial waste and require licensed hospital waste contractors to manage the collection, treatment, and disposal of this waste in line with environmental standards. Color-coded disposal bags are used for different waste types, including yellow for biohazardous, purple for cytotoxic, red for radioactive, and black for general waste. Additionally, specialized disposal procedures are in place for used syringes and pharmaceutical waste to ensure safe handling, with (present) six designated waste disposal contractors utilizing enclosed trucks and incinerators to handle biohazardous and cytotoxic waste from healthcare facilities. These facilities utilize innovative technologies, such as microwave pyrolysis, to efficiently convert BMW into syngas. This not only helps manage waste but also contributes to Singapore's energy needs and environmental objectives (Tan & Khoo, 2006).
- In 2021, *Sweden* achieved a significant milestone in BMW recycling, surpassing Singapore with nearly 95% recycling. This feat is especially remarkable given Sweden's substantial BMW output from its well-established healthcare system. Sweden's success is primarily attributed to the government's unwavering commitment to environmental sustainability, supported by key legislation like the Producer Responsibility Act, waste ordinance, and environmental code. The nation's strong culture of environmental responsibility, evident within both corporations and individuals, has played a pivotal role. Swedish businesses are increasingly adopting CE practices, while citizens actively engage in recycling and ecological initiatives. An exemplary instance is Avfall Sverige, a nonprofit organization operating various biological waste treatment facilities. Their management process begins with the categorization of waste into separate groups, considering the

varying levels of risk involved, thereby ensuring the secure and proper handling of the waste. Licensed waste collectors specifically gather medical waste from hospitals and clinics, employing dedicated vehicles for transportation to authorized waste treatment facilities. At these facilities, diverse methods such as chemical disinfection, autoclaving, and controlled burning are employed to manage the medical waste effectively. Finally, the last stage involves disposal, where licensed landfills are utilized to safely dispose of the treated medical waste, ensuring compliance with stringent regulations and safety standards throughout the process. They use innovative recycling methods like microwave pyrolysis, transforming BMW into syngas for energy, alongside autoclaving and incineration. Avfall Sverige also provides waste minimization assessments and trash segregation training. Sweden's success in biomedical recycling results in reduced greenhouse gas emissions, less landfill waste, improved healthcare worker safety, and cost-effective waste management (Miftakhova & Ben Mahmoud, 2023; Wheeler, 2013). Other nations can draw inspiration from Sweden's legal framework and practices to enhance their BMW management, reduce their environmental impact, and elevate safety standards.

- *Switzerland* boasts a robust BMW management system, meticulously overseen by the Federal Office for the Environment (FOEN) through the ordinance on the prevention and disposal of waste (OPD). Mandated by this regulatory framework, all BMW generators in the country must register with the FOEN, securing a waste disposal permit and adhering to stringent guidelines. Proper segregation and storage practices are prerequisites for generators, who entrust licensed waste contractors with the collection of BMW.
- These waste contractors play a pivotal role in the transportation and treatment of BMW, strictly following FOEN's regulations, with common methods including incineration and autoclaving. To ensure compliance, regular inspections are conducted by the FOEN on both waste generators and contractors. This comprehensive approach, integrated into Switzerland's waste management infrastructure, serves to safeguard public health and environmental integrity. In 2021 alone, the system successfully treated and disposed of over 13,000 tons of BMW, exemplifying its efficacy and commitment to responsible waste management practices.
- *Denmark* sets a remarkable example of successful collaboration between the public and private sectors in BMW recycling. In 2020, the Danish Environmental Protection Agency (DEPA) partnered with Steno Recycling, a Danish BMW management company, to establish a cutting-edge biological waste treatment plant employing microwave pyrolysis technology. This state-of-the-art facility employs high-frequency microwaves to efficiently convert biological waste into syngas, a versatile synthetic gas used for various applications, including product manufacturing and electricity generation (Hjelgaard, 2013). It boasts a daily production capacity of 2.5 megawatts of electricity and processes up to 30 tons of biological waste. Surplus electricity is supplied to the grid, ensuring energy balance.

Denmark's BMW recycling program offers several advantages. It significantly reduces landfill dumping and incineration, saving energy and curbing emissions. It also enhances healthcare worker safety by minimizing exposure to hazardous materials and improves cost-efficiency by reducing expensive disposal methods. Denmark's accomplishments serve as a model for other countries, offering insights to mitigate environmental impacts, enhance safety standards, and bolster the financial viability of BMW management through legislative and programmatic emulation.

Numerous innovative disposal and recycling technologies are globally evolving amid the escalating waste crisis. More solutions will emerge as the crisis grows. Key lessons and best practices from these case studies include:

- *Government support proves crucial*: Implementing advanced disposal and recycling methods thrives with government backing, through subsidies, tax benefits, or incentives.
- *Public-private alliances yield results*: Many pioneering waste management technologies stem from collaborations between the public and private sectors. This taps into private sector expertise and resources.
- *Community involvement is vital*: To succeed, innovative disposal and recycling practices must involve the community in planning and execution. This fosters public acceptance and support for these technologies (Fig. 1.5).

1.10 Future directions and recommendations

The COVID-19 pandemic has sparked a notable surge in medical waste generation within hospitals, rendering traditional disposal methods unsustainable. Managing BMW has become a critical global challenge, necessitating continuous innovation, collaboration, and sustainable solutions. In the realm of BMW disposal and recycling technologies, several key directions and collaborative efforts are emerging.

The future will prioritize enhancing **waste-to-energy technologies**, making them more efficient, eco-friendly, and economically viable. These technologies aim to convert medical waste into energy while minimizing harmful emissions. Collaborative research among governments, industry leaders, and environmental organizations is essential for driving innovation in this domain. Among the technologies for converting medical waste into energy, thermochemical treatment stands out. It involves heating waste in an oxygen-deprived environment, known as gasification, to produce gas for power generation (Yadav, 2023). Converting bio-waste to municipal waste is a promising approach, reducing the volume and weight of BMW through grinding, heating, and sterilization. The outcome is inert, compact, and dry waste suitable for use as refuse-derived fuel (RDF) (Fig. 1.6) Co., S. 2017.

Recycling BMW faces challenges related to infection transmission and improper classification of mixed plastics (Rahimi & García, 2017; Walker, 2021). Also, green chemistry and chemical neutralization processes are being developed to make pharmaceutical waste treatment safer and more sustainable through cooperation between researchers, chemical engineers, and regulatory bodies.

The future of BMW management will embrace **CE** models, involving collaboration between healthcare manufacturers, recyclers, and regulators to create sustainable,

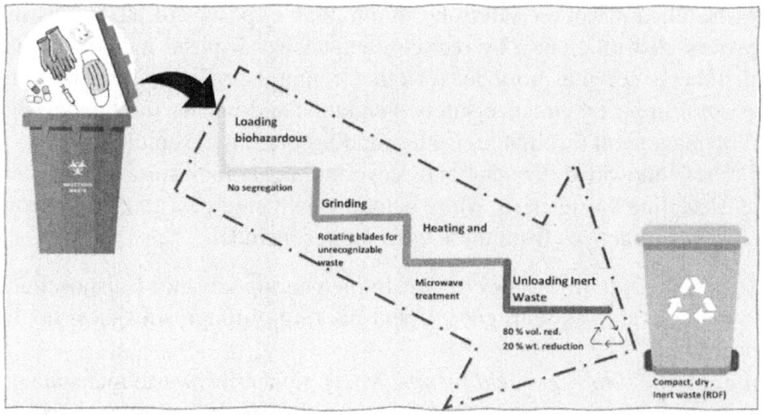

FIGURE 1.6

Conversion of BMW to RDF process. *BMW,* biomedical waste; *RDF,* refuse-derived fuel.

reusable, and recyclable healthcare products to reduce pharmaceutical waste and environmental contamination risks.

Raising public awareness about proper BMW disposal is an ongoing effort, requiring collaboration between governments, healthcare institutions, and environmental organizations to inform communities.

Digital tracking and monitoring systems, leveraging the IoT and data analytics, provide real-time information on waste streams and optimize waste management. IoT facilitates communication between nondigital entities, enhancing the precision of waste collection through real-time tracking of bin status and waste levels (Shirke et al., 2008).

Finally, **global as well as cross-sector collaboration** between countries and international organizations is crucial for adopting best practices and standards in BMW management. Continued research and development in BMW management technologies are vital, necessitating collaboration between research institutions, government funding agencies, and industry partners.

In conclusion, the future of BMW management relies on sustainability, innovation, and collaboration. These emerging strategies have the potential to lessen environmental harm from medical waste, safeguard public health, and preserve resources. Moving forward, it is crucial for stakeholders across sectors to unite, forging a sustainable and responsible approach to BMW management.

References

Adnan, Ong, Nomanbhay, Chew, & Show (2019). Technologies for biogas upgrading to biomethane: A review. *Bioengineering, 6*(4), 92. doi:10.3390/bioengineering6040092.

Aithal, P. S., et al. (2021). Smart city waste management through ICT and IoT driven solution. *International Journal of Applied Engineering and Management Letters (IJAEML), 5*(1), 51–65.

Ali, M., Wang, W., Chaudhry, N., & Geng, Y. (2017). Hospital waste management in developing countries: A mini review. *Waste Management and Research, 35*(6), 581–592. doi:10.1177/0734242X17691344.

Asim, N., Badiei, M., & Sopian, K. (2021). Review of the valorization options for the proper disposal of face masks during the COVID-19 pandemic. *Environmental Technology and Innovation*, 23. doi:10.1016/j.eti.2021.101797.

Attrah, M., Elmanadely, A., Akter, D., & Rene, E. R. (2022). A review on medical waste management: Treatment, recycling, and disposal options. *Environments - MDPI, 9*(11). doi:10.3390/environments9110146.

Beheshtinia, M. A., Bahrami, F., Fathi, M., & Asadi, S. (2023). Evaluating and prioritizing the healthcare waste disposal center locations using a hybrid multi-criteria decision-making method. *Scientific Reports, 13*(1), 1–16. doi:10.1038/s41598-023-42455-w.

Blenkharn, J. I. (2009). Sharps management and the disposal of clinical waste. *British Journal of Nursing, 18*(14), 860–864. doi:10.12968/bjon.2009.18.14.43353.

Challenge, W.H.O.G.P.S. (2017). *Medication without harm.* World Health Organization.

Chauhan, A., Jakhar, S. K., & Chauhan, C. (2021). The interplay of circular economy with industry 4.0 enabled smart city drivers of healthcare waste disposal. *Journal of Cleaner Production, 279*, 123854. doi:10.1016/j.jclepro.2020.123854.

Chitnis, V., Vaidya, K., & Chitnis, D. (2005). Biomedical waste in laboratory medicine: Audit and management. *Indian Journal of Medical Microbiology, 23*(1), 6–13. doi:10.1016/S0255-0857(21)02704-3.

Chu, Y. T., Zhou, J., Wang, Y., Liu, Y., & Ren, J. (2023). Current state, development and future directions of medical waste valorization. *Energies, 16*(3). doi:10.3390/en16031074.

Co.. Innovative Solution for Biomedical Waste Management Biomedical Waste : a World-wide Issue for. Public Health & Environment. https://www.bertin-medical-waste.com/wp-content/uploads/2017/08/Plaquette_Sterilwave_EN_web.pdf.

Cohoon, B. D. (2002). Reprocessing single-use medical devices. *AORN Journal, 75*(3), 557–567. doi:10.1016/S0001-2092(06)61177-0.

Czerwińska, K., Śliz, M., & Wilk, M. (2022). Hydrothermal carbonization process: Fundamentals, main parameter characteristics and possible applications including an effective method of SARS-CoV-2 mitigation in sewage sludge. A review. *Renewable and Sustainable Energy Reviews, 154*, 111873. doi:10.1016/j.rser.2021.111873.

Das, S. K., Chinnappan, A., Jayathilaka, W. A. D. M., Gosh, R., Baskar, C., & Ramakrishna, S. (2021). Challenges and potential solutions for 100% recycling of medical textiles. *Materials Circular Economy, 3*(1), 13. doi:10.1007/s42824-021-00023-5.

Datta, P., Mohi, G., & Chander, J. (2018). Biomedical waste management in India: Critical appraisal. *Journal of Laboratory Physicians, 10*(01), 006–014. doi:10.4103/jlp.jlp_89_17.

Demirbas, A. (2011). Waste management, waste resource facilities and waste conversion processes. *Energy Conversion and Management, 52*(2), 1280–1287. doi:10.1016/j.enconman.2010.09.025.

Dharmaraj, S., Ashokkumar, V., Pandiyan, R., Halimatul Munawaroh, H. S., Chew, K. W., Chen, W.-H., et al. (2021). Pyrolysis: An effective technique for degradation of COVID-19 medical wastes. *Chemosphere, 275*, 130092. doi:10.1016/j.chemosphere.2021.130092.

Dri, M., Canfora, P., Antonopoulos, I. S., & Gaudillat, P. (2018). Best environmental management practice for the waste management sector. *JRC Science for Policy Report*. doi:10.2760/50247.

El-Ramady, H., Brevik, E. C., Elbasiouny, H., Elbehiry, F., Amer, M., Elsakhawy, T., et al. (2021). Planning for disposal of COVID-19 pandemic wastes in developing countries: A review of current challenges. *Environmental Monitoring and Assessment*, (9), 193. doi:10.1007/s10661-021-09350-1.

Erdogan, A. A., & Yilmazoglu, M. Z. (2021). Plasma gasification of the medical waste. *International Journal of Hydrogen Energy, 46*(57), 29108–29125. doi:10.1016/J.IJHYDENE.2020.12.069.

Esmaeilian, B., Wang, B., Lewis, K., Duarte, F., Ratti, C., & Behdad, S. (2018). The future of waste management in smart and sustainable cities: A review and concept paper. *Waste Management, 81*, 177–195. doi:10.1016/j.wasman.2018.09.047.

Fakayode, O. A., Aboagarib, E. A. A., Zhou, C., & Ma, H. (2020). Co-pyrolysis of lignocellulosic and macroalgae biomasses for the production of biochar – A review. *Bioresource Technology, 297*, 122408. doi:10.1016/J.BIORTECH.2019.122408.

Fatimah, Y. A., Widianto, A., & Hanafi, M. (2020). Cyber-physical system enabled in sustainable waste management 4.0: A smart waste collection system for indonesian semi-urban cities. *Procedia Manufacturing, 43*, 535–542. doi:10.1016/j.promfg.2020.02.169.

Gautam, S., Bhatnagar, D., Bansal, D., Batra, H., & Goyal, N. (2022). Recent advancements in nanomaterials for biomedical implants. *Biomedical Engineering Advances, 3*, 100029. doi:10.1016/j.bea.2022.100029.

Gautam, V., Thapar, R., & Sharma, M. (2010). Biomedical waste management: Incineration vs. environmental safety. *Indian Journal of Medical Microbiology, 28*(3), 191–192. doi:10.4103/0255-0857.66465.

Ghasemi, M. K., & Yusuff, R. B. M. (2016). Advantages and disadvantages of healthcare waste treatment and disposal alternatives: Malaysian scenario. *Polish Journal of Environmental Studies, 25*(1), 17–25. doi:10.15244/pjoes/59322.

Gomez, E., Rani, D. A., Cheeseman, C. R., Deegan, D., Wise, M., & Boccaccini, A. R. (2009). Thermal plasma technology for the treatment of wastes: A critical review. *Journal of Hazardous Materials, 161*(2–3), 614–626. doi:10.1016/j.jhazmat.2008.04.017.

Guzzo, D., Carvalho, M. M., Balkenende, R., & Mascarenhas, J. (2020). Circular business models in the medical device industry: Paths towards sustainable healthcare. *Resources, Conservation and Recycling, 160*, 104904. doi:10.1016/j.resconrec.2020.104904.

Hart, J. (2022). Health care without harm: Taking environmental action to improve lives and the planet. *Integrative and Complementary Therapies, 28*(5), 251–254. doi:10.1089/ict.2022.29036.pro.

Hazardous Waste, Experts. (2020). *Medical Waste Disposal Methods* https://www.hazardouswasteexperts.com/medical-waste-disposal-methods/.

Heberlein, J., & Murphy, A. B. (2008). Thermal plasma waste treatment. *Journal of Physics D: Applied Physics, 41*(5), 053001. doi:10.1088/0022-3727/41/5/053001.

Hjelgaard, K. (2013). Danish Emission Inventory for Waste Incineration and Other Waste. *and no.: Scientific Report from DCE–Danish Centre for Environment and Energy*, (70).

Jacob, S., Nithianandam, S., Rastogi, S., Sakhuja, S., & Sri Laxma Alankar, S. N. (2021). Handling and treatment strategies of biomedical wastes and biosolids contaminated with SARS-CoV-2 in waste environment. *Environmental and Health Management of Novel Coronavirus Disease (COVID-19*, 207–232. doi:10.1016/B978-0-323-85780-2.00012-3.

Jain, M., Sai Kiran, P., Ghosal, P. S., & Gupta, A. K. (2023). Development of microbial fuel cell integrated constructed wetland (CMFC) for removal of paracetamol and diclofenac in hospital wastewater. *Journal of Environmental Management, 344*, 118686. doi:10.1016/j.jenvman.2023.118686.

Joseph, B., James, J., Kalarikkal, N., & Thomas, S. (2021). Recycling of medical plastics. *Advanced Industrial and Engineering Polymer Research, 4*(3), 199–208. doi:10.1016/j.aiepr.2021.06.003.

Kadam, P.S., Rajagopal, N.K., Yadav, A.K., Madduri, A., Ansari, M.J., Patil, P.Y. Biomedical waste management during pandemics; 2023; p 050004. doi:10.1063/5.0126273.

Kannan, D., Khademolqorani, S., Janatyan, N., & Alavi, S. (2024). Smart waste management 4.0: The transition from a systematic review to an integrated framework. *Waste Management, 174*, 1–14. doi:10.1016/j.wasman.2023.08.041.

Kaur, J. (2023). Advances in biomedical waste management technologies. In *Waste management and resource recycling in the developing world* (pp. 543–573). Elsevier. doi:10.1016/B978-0-323-90463-6.00024-5.

Kaya, M. (2016). Recovery of metals and nonmetals from electronic waste by physical and chemical recycling processes. *Waste Management, 57*, 64–90. doi:10.1016/j.wasman.2016.08.004.

Kheirabadi, S., & Sheikhi, A. (2022). Recent advances and challenges in recycling and reusing biomedical materials. *Current Opinion in Green and Sustainable Chemistry, 38*, 100695. doi:10.1016/j.cogsc.2022.100695.

Kollu, V. K. R., Kumar, P., & Gautam, K. (2022). Comparison of microwave and autoclave treatment for biomedical waste disinfection. *Systems Microbiology and Biomanufacturing, 2*(4), 732–742. doi:10.1007/s43393-022-00101-y.

Lalman, C., Karunathilake, H., & Ruparathna, R. (2023). To dispose or to reuse? analyzing the life cycle impacts and costs of disposal, sterilization, and reuse of electrophysiological catheters. *Sustainability, 15*(6), 5363. doi:10.3390/su15065363.

Lamba, P., Kaur, D. P., Raj, S., & Sorout, J. (2022). Recycling/reuse of plastic waste as construction material for sustainable development: A review. *Environmental Science and Pollution Research, 29*(57), 86156–86179. doi:10.1007/s11356-021-16980-y.

Likotiko, E. D., Nyambo, D., & Mwangoka, J. (2017). Multi-agent based iot smart waste monitoring and collection architecture. *International Journal of Computer Science, Engineering and Information Technology, 7*(5), 01–14. doi:10.5121/ijcseit.2017.7501.

Lyu, L., Bagchi, M., Markoglou, N., An, C., Peng, H., Bi, H., et al. (2024). Towards environmentally sustainable management: A review on the generation, degradation, and recycling of polypropylene face mask waste. *Journal of Hazardous Materials, 461*, 132566. doi:10.1016/j.jhazmat.2023.132566.

Manegdeg, F., Coronado, L. O., & Paña, R. (2020). Medical waste treatment and electricity generation using pyrolyzer-rankine cycle for specialty hospitals in Quezon City, Philippines. *IOP Conference Series: Earth and Environmental Science, 463*(1), 012180. doi:10.1088/1755-1315/463/1/012180.

Mang, B., Oh, Y., Bonilla, C., & Orth, J. (2023). A medical equipment lifecycle framework to improve healthcare policy and sustainability. *Challenges, 14*(2), 21. doi:10.3390/challe14020021.

Martínez-Barbosa, M. E., & Moreno-Corral, R. A. (2022). Washable, reusable and disposable medical textiles. In *Medical Textiles from Natural Resources* (pp. 717–765). Elsevier. doi:10.1016/B978-0-323-90479-7.00017-8.

Mastellone, M. L., & Zaccariello, L. (2010). Arena, U. Co-gasification of coal, plastic waste and wood in a bubbling fluidized bed reactor. *Fuel, 89*(10), 2991–3000. doi:10.1016/J.FUEL.2010.05.019.

Miftakhova, A., & Ben Mahmoud, C.. *Reconciling material & social through waste: The case of överskottet as an alternative way of organizing waste*. https://www.diva-portal.org/smash/get/diva2:1776502/FULLTEXT01.pdf.

Mohamed, N. H., Khan, S., & Jagtap, S. (2023). Modernizing medical waste management: Unleashing the power of the internet of things (IoT). *Sustainability (Switzerland), 15*(13), 1–21. doi:10.3390/su15139909.

Newman, S. (2023). What a waste ! an evaluation of federal and state medical and biohazard waste regulations during the COVID-19 pandemic and their impact on environmental justice. *Vill. Env't LJ, 34*(1), 57.

Ojha, P. C., Satpathy, S. S., Ojha, A. K., Sukla, L. B., & Pradhan, D. (2022). Overcoming challenges due to enhanced biomedical waste generation during COVID-19 pandemic. *Science of The Total Environment, 832*, 155072. doi:10.1016/j.scitotenv.2022.155072.

Ordway, A., Pitonyak, J. S., & Johnson, K. L. (2020). Durable medical equipment reuse and recycling: Uncovering hidden opportunities for reducing medical waste. *Disability and Rehabilitation: Assistive Technology, 15*(1), 21–28. doi:10.1080/17483107.2018.1508516.

Paul, S., Sen, B., Das, S., Abbas, S. J., Pradhan, S. N., Sen, K., et al. (2021). Incarnation of bioplastics: Recuperation of plastic pollution. *International Journal of Environmental Analytical Chemistry*, 1–24. doi:10.1080/03067319.2021.1983552.

Perkins, D. N., Brune Drisse, M. N., Nxele, T., & Sly, P. D. (2014). E-waste: A global hazard. *Annals of Global Health, 80*(4), 286–295. doi:10.1016/j.aogh.2014.10.001.

Piasecki, J., Walkiewicz-Żarek, E., Figas-Skrzypulec, J., Kordecka, A., & Dranseika, V. (2021). Ethical issues in biomedical research using electronic health records: A systematic review. *Medicine, Health Care and Philosophy, 24*(4), 633–658. doi:10.1007/s11019-021-10031-6.

Prem Ananth, A., Prashanthini, V., & Visvanathan, C. (2010). Healthcare waste management in Asia. *Waste Management, 30*(1), 154–161. doi:10.1016/J.WASMAN.2009.07.018.

Rahimi, A., & García, J. M. (2017). Chemical recycling of waste plastics for new materials production. *Nature Reviews Chemistry, 1*(6), 46. doi:10.1038/s41570-017-0046.

Rajan, R., Robin, D. T., & Vandanarani, M. (2019). Biomedical waste management in Ayurveda hospitals – Current practices and future prospectives. *Journal of Ayurveda and Integrative Medicine, 10*(3), 214–221. doi:10.1016/j.jaim.2017.07.011.

Randhawa, G. K., & Kullar, J. S. (2011). Bioremediation of pharmaceuticals, pesticides, and petrochemicals with Gomeya/cow dung. *ISRN Pharmacology, 2011*, 1–7. doi:10.5402/2011/362459.

Reaching Out. (2011). *Orthopaedic Nursing, 30*(6), 351–352. doi:10.1097/NOR.0b013e31823d5095.

Rowan, N. J., & Laffey, J. G. (2021). Unlocking the surge in demand for personal and protective equipment (PPE) and improvised face coverings arising from coronavirus disease (COVID-19) pandemic – Implications for efficacy, re-use and sustainable waste management. *Science of The Total Environment, 752*, 142259. doi:10.1016/j.scitotenv.2020.142259.

Rutala, W. A., & Weber, D. J. (2015). Disinfection, sterilization, and control of hospital waste. In *Mandell, Douglas, and Bennett's principles and practice of infectious diseases* (pp. 3294–3309.e4). Elsevier. doi:10.1016/B978-1-4557-4801-3.00301-5.

Sahoo, S., Rathod, W., Vardikar, H., Biswal, M., Mohanty, S., & Nayak, S. K. (2023). Biomedical waste plastic: Bacteria, disinfection and recycling technologies—A comprehensive review. *International Journal of Environmental Science and Technology*. doi:10.1007/s13762-023-04975-w.

Sapkota, B., & Pariatamby, A. (2023). Pharmaceutical waste management system – Are the current techniques sustainable, eco-friendly and circular? A review. *Waste Management, 168*, 83–97. doi:10.1016/j.wasman.2023.05.052.

Sharma, H. B., Vanapalli, K. R., Cheela, V. S., Ranjan, V. P., Jaglan, A. K., Dubey, B., et al. (2020). Challenges, opportunities, and innovations for effective solid waste management during and post COVID-19 pandemic. *Resources, Conservation and Recycling*, 162. doi:10.1016/j.resconrec.2020.105052.

Sharma, S., & Dubey, A. (2022). Advances in bioremediation of biosurfactants and biomedical wastes. In *Biological Approaches to Controlling Pollutants* (pp. 259–272). Elsevier. doi:10.1016/B978-0-12-824316-9.00011-2.

Shen, Y., Yu, S., Ge, S., Chen, X., Ge, X., & Chen, M. (2017). Hydrothermal carbonization of medical wastes and lignocellulosic biomass for solid fuel production from lab-scale to pilot-scale. *Energy, 118*, 312–323. doi:10.1016/j.energy.2016.12.047.

Shirke, S. I., Ithape, S., Lungase, S., & Mohare, M. (2008). Automation of smart waste management using IoT. *International Research Journal of Engineering and Technology*, 414.

Shukla, S., Kalaiselvan, V., & Singh Raghuvanshi, R. (2023). How to improve regulatory practices for refurbished medical devices. *Bulletin of the World Health Organization, 101*(06), 412–417. doi:10.2471/BLT.22.289416.

Singh, E., Kumar, A., Mishra, R., & Kumar, S. (2022). Solid waste management during COVID-19 pandemic: Recovery techniques and responses. *Chemosphere, 288*. doi:10.1016/j.chemosphere.2021.132451.

Singh, Z., Bhalwar, R., Jayaram, J., & Tilak, V. W. (2001). An introduction to essentials of bio-medical waste management. *Medical Journal Armed Forces India, 57*(2), 144–147. doi:10.1016/S0377-1237(01)80136-2.

Sivadas, S. K., Mishra, P., Kaviarasan, T., Sambandam, M., Dhineka, K., Murthy, M. V. R., et al. (2022). Litter and plastic monitoring in the Indian marine environment: A review of current research, policies, waste management, and a roadmap for multidisciplinary action. *Marine Pollution Bulletin, 176*, 113424. doi:10.1016/j.marpolbul.2022.113424.

Subramanian, A. K., Thayalan, D., Edwards, A. I., Almalki, A., & Venugopal, A. (2021). Biomedical waste management in dental practice and its significant environmental impact: A perspective. *Environmental Technology & Innovation, 24*, 101807. doi:10.1016/j.eti.2021.101807.

Tan, R. B. H., & Khoo, H. H. (2006). Impact assessment of waste management options in Singapore. *Journal of the Air and Waste Management Association, 56*(3), 244–254. doi:10.1080/10473289.2006.10464463.

Thakur, M., Chandel, M., Rani, A., Sharma, A., & Pathania, D. (2023). Chemical methods for the treatment of biomedical hazardous waste. In *Waste Management and Resource Recycling in the Developing World* (pp. 521–541). Elsevier. doi:10.1016/B978-0-323-90463-6.00008-7.

Thakur, Y., & Katoch, S. S. (2012). Emerging technologies in biomedical waste treatment and disposal. *Chemical Engineering Transactions, 29*, 787–792. doi:10.3303/CET1229132.

van Straten, B., Dankelman, J., van der Eijk, A., & Horeman, T. (2021). A circular healthcare economy; a feasibility study to reduce surgical stainless steel waste. *Sustainable Production and Consumption, 27*, 169–175. doi:10.1016/j.spc.2020.10.030.

Walker, T. R. (2021). (Micro)plastics and the UN sustainable development goals. *Current Opinion in Green and Sustainable Chemistry, 30*, 100497. doi:10.1016/j.cogsc.2021.100497.

Wang, J., Shen, J., Ye, D., Yan, X., Zhang, Y., Yang, W., et al. (2020). Disinfection technology of hospital wastes and wastewater: Suggestions for disinfection strategy during coronavirus Disease 2019 (COVID-19) pandemic in China. *Environmental Pollution, 262*, 114665. doi:10.1016/j.envpol.2020.114665.

Wheeler, K. (2013). *The largest environmental movement: Recycling and consumption work in Sweden* https://www.semanticscholar.org/paper/The-Largest-Environmental-Movement%3A-Recycling-and-Wheeler/b69f3100d0c381c8fb8e4b4a5289819e41499111.

Xu, B.-X., Liu, Z., & Rustam, A. (2023). Application of extended theory of planned behavior to explore household pharmaceutical waste recycling intentions: A case study of China. *Journal of Material Cycles and Waste Management, 25*(5), 2870–2886. doi:10.1007/s10163-023-01721-9.

Yadav, P. (2023). Environmental biotechnology for medical waste management: A review of current practices and future directions. *Annals of Advanced Biomedical Sciences, 6*(2). doi:10.23880/aabsc-16000214.

Zhang, F., Zhao, Y., Wang, D., Yan, M., Zhang, J., Zhang, P., et al. (2021). Current technologies for plastic waste treatment: A review. *Journal of Cleaner Production, 282*, 124523. doi:10.1016/j.jclepro.2020.124523.

Zhao, X., Chen, H., Zheng, Q., Liu, J., Pan, P., Xu, G., et al. (2023). Thermo-economic analysis of a novel hydrogen production system using medical waste and biogas with zero carbon emission. *Energy, 265*, 126333. doi:10.1016/j.energy.2022.126333.

Zimmermann, K. (2017). Microwave as an emerging technology for the treatment of biohazardous waste: A mini-review. *Waste Management & Research: The Journal for a Sustainable Circular Economy, 35*(5), 471–479. doi:10.1177/0734242X16684385.

CHAPTER

2 Impact of the COVID-19 pandemic on biomedical waste generation and management

Ramkrishna Mondal

Department of Hospital Administration, All India Institute of Medical Sciences Patna, Patna, Bihar, India

2.1 Introduction

Improper waste disposal, whether it is medical waste, healthcare waste, biomedical waste (BMW), household waste, or municipal solid waste (MSW), has serious public health and environmental consequences. Proper waste management is required to avoid any negative effects on the environment and health. Coronavirus disease (COVID-19) creates additional waste management issues, and the need for effective waste management should be highlighted during this pandemic. COVID-19 moved quickly throughout China, exposing BMW management's flaws. As a result of the high infection rate, COVID-19 was designated an International Public Health Emergency by the WHO on January 31, 2020, as it swiftly spread in China, followed by the rest of the world (Bashir et al., 2020). In Wuhan, the daily BMW generation increased fivefold, from 40 tons to 250 tons, significantly surpassing the daily processing capability of the BMW treatment facility. Until the first 2 months of 2020, China had the highest number of new cases and deaths, while neighboring nations such as Asia and North America had lower numbers (Bashir et al., 2020a), and Europe had been relatively safe (Chakraborty & Maity, 2020). However, all of them suffered greatly later, and some of them are now in the second wave phase. During COVID-19, both China and the United States raised similar issues regarding BMW management, such as waste management. According to the WHO report, there is little evidence of COVID transmission while handling BMW by waste handlers who do not properly protect themselves (WHO & UNICEF, 2020). According to the Solid Waste Association of North America, sanitation workers who maintain social distances from other people are less likely to become infected with the virus (Calma, 2020). Personal protective equipment (PPE) was in high demand worldwide to protect doctors, nurses, healthcare workers, and waste handling personnel, resulting in massive PPE production. As a result of lockdowns and the implementation of home quarantines to control infection, waste generation has also increased, whether infected or not,

Biomedical Waste Management in the COVID-19 Era: A Global Scenario.
DOI: https://doi.org/10.1016/B978-0-443-21841-5.00002-3

during COVID-19 (Lal et al., 2020). Although the pandemic had a negative impact on global economic activity, it also resulted in numerous positive environmental changes (Espejo et al., 2020). Not only has the pandemic had a significant impact on BMW generation, but also on MSW production and composition. According to the WHO and the US CDC, pandemic waste can be divided into two categories: (1) within medical facilities and (2) outside of medical facilities (Somani et al., 2020). COVID-19 can be said to have altered the waste production dynamic (Ragazzi et al., 2020) with potential criticality in all stages of waste management, beginning with segregation, collection, transportation, treatment, and disposal (Peng et al., 2020). To maintain social distance and other measures to protect against COVID-19 during the pandemic, there is an increase in plastic use, a situation that requires policy consequences (Klemeš et al., 2020). This increased use of plastic creates a channel for contamination and increases the likelihood of spread. To control the pandemic, a movement ban was imposed, affecting countries that rely on external technologies for waste reprocessing and management. As a result, waste generated during that time is discarded rather than recycled. COVID-19 pandemic-like emergencies have made BMW's long-term management difficult and complicated. The bulky waste must be collected for disposal or recycled. Both processes are jeopardized due to a lack of manpower and the difficulty of implementing infection control measures (Association of Cities and Regions for Sustainable Resource Management, 2020; European Commission, 2020). These disruptions in services result in waste mishandling, which has increased by up to 300% in a few rural communities in the United Kingdom (UK). Because of the pandemic, the UK Environment Agency permitted the temporary storage of waste and incineration ash at sites that endanger the environment (UK Environment Agency, 2020). To reduce the socioeconomic and environmental impacts of waste, the waste management system must be considered, from waste generation to treatment, including resource use, and disposal of excess waste. Shielding waste management chains will also help to achieve sustainable cities and communities (United Nations, 2020). The COVID-19 pandemic is the third major zoonotic disease outbreak in the last two decades. Severe acute respiratory syndrome (SARS) struck in 2002–2003, followed by Middle East Respiratory Syndrome (MERS) in 2012. On December 31, 2019, Chinese health officials reported the disease to the WHO as an atypical pneumonia for unknown reasons (Poon & Peiris, 2020). According to van Doremalen et al. (2020), depending on the materials and environment, the survival period can range from hours to days. Survival phases on copper and cardboard range from 3 to 4 h, while on plastic and stainless steel, survival phases range from 1 to 3 days, and in some cases up to 9 days, even in tap or wastewater at 20°C for 2 days (Wang et al., 2020). Because of the longer survival period, there is a high risk of communal spread by inappropriate use. This chapter first discusses various impacts, such as socioeconomic, environmental, and impacts on waste generation and waste management. Following that, various challenges encountered due to these impacts on waste management are listed. Finally, it described the new models or rethinking of waste management processes during a pandemic.

2.2 Impact on socioeconomic and environmental

COVID-19 is distressing everyone differently, according to socioeconomic and demographic data (Collivignarelli et al., 2020). Due to a lack of suitable data, it is extremely difficult to fully understand why this COVID-19 affects different groups of people differently in society, with the poor suffering the most (Messner, 2020). Furthermore, the pandemic caused significant disruption in both local and global business (Goodell & Huynh, 2020). Even after corrective and preventive measures were implemented, it was discovered that very few international communications began in China (Bogoch et al., 2020). Different economic institutes and the World Bank are very concerned about the economic impact of a pandemic. To mitigate the socioeconomic impact, job and business losses must be restored, as well as the slowing of national transformations. Following the outbreak of the pandemic, stock exchange markets experienced a significant drop following the 2008 economic disaster. The market slide continues, with global markets also falling significantly. However, it will gradually increase after October/November 2020. As a global pandemic, COVID-19 caused numerous environmental issues, particularly in BMW management. Wuhan generates 200 tons of BMW, which is four times the capacity of its waste treatment facilities (Luan & Ching, 2020). Ocean Asia, a leading NGO, stated that BMW produced in Hong Kong harmed the seaside environment (Hellewell et al., 2020). Countries must announce lockdown measures in response to COVID-19 to protect their citizens and limit their exposure (Zambrano-Monserrate & Ruano, 2020). Several air changes have been discovered around the world, such as a 50% reduction in air pollutants in New York, a 25% reduction in emissions in China due to a 40% reduction in coal use, and a significant reduction in air pollution during a pandemic (Gautam, 2020). Even though the financial impacts of COVID-19 were significant, it significantly contributed to an eco-friendly, decarbonized environment, and air quality improved significantly, as reported in the various literature (Conticini et al., 2020; Sharif et al., 2020). NASA satellite images used to analyze air quality show a 20%–38% reduction in pollutants in Korea, Spain, Germany, and the United States (Knowland et al., 2022).

2.3 Impact on waste generation

BMW generated from the COVID-19 patient wards is also increasing and becoming a bigger concern for the hospital as the number of patients increases. As a result, we must ensure that contaminated hospital trash does not endanger the public before it is safely disposed of. BMW characteristics have changed: BMW includes waste generated by hospitals, medical institutions, medical laboratories, and other facilities where patients are cared for (Rupani et al., 2020). In general, 75% to 90% of hospital waste is general, nonhazardous waste that does not require special management, and only 10%–25% is BMW that requires special disposal. In a pandemic, however, all wastes generated within a facility would be measured as BMW and managed with extreme caution. To manage this unusual amount of BMW, existing waste treatment facilities must be modified based on actual production data and treatment facility availability (Sarkodie & Owusu, 2020a). According to a UNEP report, hospitals that generated 0.5 kg/bed/day of waste before the pandemic generate 3.4 kg/bed/day during the pandemic. According to study (Ramteke & Sahu, 2020) waste generation in India increased from 0.5 to 2.54 kg/bed/day (Ramteke & Sahu, 2020). In addition, in a study conducted at a Chinese hospital by Peng et al. (2020). During the pandemic, the average waste generation rate was more than 6 kg/bed/day. The COVID-19 virus causes asymptomatic infections (Mizumoto et al., 2020; Nishiura et al., 2020). As a result, determining the level of COVID-19 transmission in communities and between countries is difficult. COVID-19 surveillance in wastewater is an effective method for detecting and analyzing virus spread because it provides unbiased results and can be used as an early warning system. Wastewater monitoring is a virus identification system for viruses that have contaminated the environment and society; reducing COVID-19 in wastewater can also be used as a marker for reducing COVID-19 cases in the community or region (O'Bannon et al., 2020). The most recent study in the United States used a wastewater control approach. It was used to determine the phylogeny of the COVID-19 strain and to assess the efficacy of public health interventions to combat outbreaks (Nemudryi et al., 2020). Waste generation in some countries and regions has decreased due to a variety of factors, such as lockdowns and travel restrictions, particularly in tourist cities such as Barcelona, Spain, by 25%, Macao, China, by 17%, 25%, and Catalonia, Spain, by 17%. COVID-19, on the other hand, causes an increase in waste generation in some areas, such as England, where waste generation is more than 90% higher than usual (ADEPT, 2020). The social distancing measures reduced BMW production while shifting some of the BMW production to MSW (Naughton, 2020). Most developing countries were dealing with issues of poor BMW management long before the pandemic (Rahman et al., 2020). According to an Asian Development Bank report, during the pandemic, Manila, Kuala Lumpur, Hanoi, and Bangkok produced 154–280 tons of BMW per day [Asian Development Bank (ADB), 2020].

The trends of global trash generation are explored and compared with the areas where MSW and biomedical waste (BMW) have decreased and increased. Major drops in MSW have occurred in Macao, Morocco, Trento, Milan, Catalonia, and Ahmedabad; increases in England, Tehran, South Korea, New York, and Tokyo. For example, MSW increased by as much as 20% in England, and MSC decreased by as much as 40% in Ahmedabad, while BMW experienced considerable growth. The data shows overall different tendencies which are affected by local laws and customs (ADEPT 2020; Mondal & Chakrabarty, 2022; Ragazzi et al., 2020; Somani et al., 2020; Zand & Heir, 2020).

2.3.1 Personal protective equipment waste

PPE is still the most dependable and reasonably priced item for preventing virus contamination and spread. PPE was initially used primarily by healthcare professionals, but as the pandemic progressed, PPE was widely used by the general population, resulting in an increase in demand for PPE during the pandemic. Almost every country requires its citizens to wear face masks. Some countries' mask use varies widely, like in the Republic of Ireland, where 9 million masks are used every week, 700 million masks are used every day across Africa, and Italy's daily use of masks is 40 million, compared with 5.5 million in Iran (Mondal & Chakrabarty, 2022; Ragazzi et al., 2020; Rodríguez-Pardo et al., 2020). A mask or PPE is mostly made of plastic (>80%) (Prata et al., 2020). The increased demand and mandatory mask use changed the dynamics of plastic waste generation, enhancing the complexities of plastic management and impeding the hard work to reduce plastic trash (Vanapalli et al., 2021).

2.3.2 Plastic waste

Plastics are being used more for convenience and reassurance as people's lifestyles change. At the same time, everyone was concerned about reducing plastic waste due to global pollution caused by plastic. During this pandemic, there is a shift from concern about plastic pollution to a preference for plastic packaging (Grodzińska-Jurczak et al., 2020). During the pandemic, demand for packaging and transportation increased dramatically (Daniel et al., 2020). To control the spread of quarantine, most countries increased demand for online shop services (Vanapalli et al., 2021). Overall, when compared to the prepandemic era, online trading has increased (16.6%) in the pandemic and postpandemic eras (Clement, 2020).

2.3.3 Food waste

COVID-19 has influenced lifestyle and consumption habits, resulting in food loss and waste (FLW). A survey of US households conducted in April 2020 discovered that during the pandemic, cooking at home increased by 54% (n = 1005) (Hunter, 2020), which implies food waste, but 57% indicated less waste, implying a net effect of no change. Individuals waste more food during a crisis because they believe there is a scarcity of food, resulting in excess food storage (Brizi & Biraglia, 2021). Spanish consumption data show an increase in food purchases (29.8% and 10.9%) in the first and second weeks of the lockdown, and then a decrease of 20.3% in the third week, indicating that the pandemic had a minimal effect on the FLW by 12% hike, which may be due to the socioeconomic crisis (Jribi et al., 2020).

2.4 Impacts on waste management

According to the Centers for Disease Control and Prevention (CDC), the risk of COVID-19 transmission is high in people who are in close contact with a diseased person or who touch a surface or object, which is a less-known mode of virus transmission. However, OSHA states that the chance of becoming acquainted will be increased for workers involved in waste and wastewater management. According to a WHO report, 21 million people were infected with Hepatitis B, 2 million with Hepatitis C, and 260,000 with HIV in 2000 because of contaminated needles (World Health Organization, 2016). However, waste thrown by isolated families may endanger waste handlers and cloth pickers. It is also critical to ensure that waste is not mixed at regular dumping sites. A good waste management strategy denotes a facility's program for managing produced waste for treatment and disposal. In response to the pandemic, each country has implemented the best management strategies available based on its resources (Singh et al., 2020; Zand & Heir, 2020). In the event of a problem with treatment or disposal methods, appropriate facilities should be available to keep waste in temporary storage in sealed containers and in highly secure areas. To avoid virus transmission, a disinfectant must be used. To dispose of the excess BMW generated by the pandemic, mobile incineration and autoclaves can be used. Waste should be transported in designated vehicles with proper documentation [Asian Development Bank (ADB) 2020] and a special permit to collect healthcare waste to ensure proper disposal. The authorities should strictly monitor the process to ensure that regulations are followed in accordance with national and local guidelines. Any good management system necessitates an assessment of available options, plans, and clearly defined personnel roles and responsibilities. A waste management committee should be formed to carry out the plan. Strategies should be reviewed on a regular basis, and all personnel involved in waste management processes should be kept up to date on any changes (Thakur, 2021). Overall, resource optimization and proper waste segregation significantly reduce waste (Rodríguez-Pardo et al., 2020). Because waste generation cannot be predicted, temporary waste treatment technologies such as autoclaves and burn incinerators are used to effectively manage waste (Yu et al., 2020). During a pandemic, proper waste management always aids in increasing the quantity of recyclable waste by utilizing these alternative technologies (Zand & Heir, 2020). Considering all these options, healthcare waste can be used to add value to the country's economy during the pandemic (Ilyas et al., 2020; Klemeš et al., 2020). Various international organizations have published numerous waste management policies and guidelines from time to time. Some countries have BMW laws, and others have issued new strategies and plans to combat the pandemic. The unexpected rise in BMW is a major challenge for everyone. India has also issued policies on BMW management from time to time (Vanapalli et al., 2021). In Iran, a pandemic causes the clearance of infectious waste to be halted for an extended period (Rodríguez-Pardo et al., 2020). The National Institute of Public Health in Romania classifies household waste as infectious waste and takes strict waste management measures (Mihai, 2020). Various decontamination and discarding methods and technologies are used depending on factors such as quantity, type, and cost (Ilyas et al., 2020; Sharif et al., 2020). In general, the most widely accepted, safe, simple, and effective technology would be a combination of chemical disinfection and incineration (Sharif et al., 2020). If BMW treatment facilities are overburdened, industrial furnaces or cement kilns can be used safely [Asian Development Bank (ADB), 2020]. On-site and mobile treatments are preferred due to their adaptability to changing demands. Fixed, mobile, and coprocessing facilities, and nonlocal disposal methods were all used sequentially in China, even though all of these are supplementary waste management techniques (Yang et al., 2021). Virus survival times on various surfaces can range from 1 to 9 days (Conticini et al., 2020), which implies that COVID patient waste could be a source of infection and that managing it will endanger waste workers (Ragazzi et al., 2020). The different countries made different adjustments in their waste management. A few common adjustments are summarized later.

2.4.1 healthcare waste/biomedical waste

Whatever best practices for safe healthcare waste management are available, including adequate human and material resources to segregate and dispose of waste safely, should be followed. All waste generated during patient care, including that from confirmed cases, is infectious and must be collected safely in accordance with national guidelines. The waste removed from the hospital should be properly handled, treated or disposed of in accordance with local laws. All healthcare waste handlers must wear appropriate PPE and practice proper hand hygiene. Given the outbreak, it is expected that the volume of infectious waste and PPE waste will increase; therefore, treatment capacity must be increased to handle this additional waste. Alternative treatment technologies, such as autoclaving or high-temperature incinerators, may be required to ensure their long-term operation. The ratio of infectious viruses in MSW is much lower than in BMW, but the risk of infection from any waste cannot be overlooked. Most countries agreed that recycling posed very little risk, and thus, normal waste handling and management could resume as before the pandemic. If the waste is contaminated with infectious waste, it must be properly managed (Ikiz et al., 2021).

2.4.2 Household waste/municipal solid waste

There are no official guidelines for household waste/MSW management, but some basic measures should be followed. A waste bag may be kept in the room for items used by the patient, and caregivers must immediately remove their gloves and masks after use and place them in another bag kept near the door while leaving. Waste bags must be closed before being removed and should never be emptied into another bag. Hand hygiene must be strictly followed when handling waste bags.

2.4.3 Special wastes like personal protective equipment waste and dead bodies

PPE waste for quarantine patients in the UK must be placed in a black bag waste bin and can be stored for 72 h. According to Patrício Silva et al. (2021), the Portuguese Environment Agency recommends that all PPE waste generated by ordinary citizens be placed in closed and watertight trash bags and disposed of preferably by incineration or landfill. China has not issued any PPE waste guidelines. According to an OCEANSASIA survey, many surgical masks were discovered lying on the Soko Islands. Although the risk of virus transmission from a dead body is very low, workers who handle dead bodies must take standard precautions. The mortuary staff preparing or handling the body must wear proper PPE, including an impermeable apron, and PPE must be removed and disposed of in accordance with protocol, as well as perform hand hygiene. A confirmed or suspected case's dead body should be properly wrapped and transferred to the mortuary area as soon as possible.

2.5 Challenges in handling waste generation and management

The spread of COVID-19 has increased the use of PPE worldwide, and all the littered PPE waste from healthcare and home may be a vector for COVID-19, due to the virus's potential ability to survive for up to 7 days on masks, gloves, and in fomites. Furthermore, as a polymer plastic, polypropylene takes a long time to degrade and is likely to result in additional plastic pollution, according to the World Health Organization (2016). A COVID-19 patient can generate 3.4 kg of waste per day, demonstrating the pandemic's gradual increase in waste quantity [Asian Development Bank (ADB), 2020]. The BMW from an Indian healthcare institution and hospital produces 517 tons per day. Only 501 tons/day of waste could be treated in 200 common biomedical management waste treatment facilities (CBMWTFs) spread across 28 Indian states. And the remaining 16 tons of waste per day are accumulated untreated every day. CBWTFs are not available in some states, including Goa, Andaman and Nicobar, Mizoram, Nagaland, and Sikkim (Vanapalli et al., 2021). Many other countries are facing similar challenges in managing massive amounts of waste [Asian Development Bank (ADB) 2020]. According to Zand and Heir (2020), Prata et al. (2020) , and Vanapalli et al. (2021), healthcare solid waste in France increased from 40% to 50%, and in the Netherlands, it increased from 40% to 50%, like in India and Iran. European governments are grappling with the challenge of retaining waste management personnel while also providing a pleasant environment and space for the additional waste generated by the pandemic. Many developing countries lack effective BMW management policies and resources. South Asia is a densely populated region of the world, accounting for 24.89% of the global population. To avoid a further health crisis, such areas must have safe and accurate BMW management. Furthermore, the urban-rural disparity in waste management is a significant challenge. According to the INCLEN Program Evaluation Network (IPEN) Study Group, a large portion of the healthcare system (82% of primary, 60% of secondary, and 54% of tertiary care) lacks a proper waste handling system (WHO & UNICEF, 2020). In rural areas, informal practitioners with very little or no practical knowledge handle BMW, and awareness of BMW segregation and management varies more among staff in urban and rural areas. Lots of studies recommended that these countries need to formulate comprehensive guidelines to seal knowledge gaps in the BMW handling.

Developing countries have neglected BMW management in obedience with the rules for a long time. The discussed issue of indiscriminate dumping of waste will pose a severe environmental impact and occupational health risks. Besides, ignorant public conduct will put waste handlers at risk as the quarantined domestic litter is disposed of with the regular household MSW. During these pandemics, like India, many countries came up with some thought-provoking problems, which include:

- The significantly increasing quantity of hazardous waste has created a difficult issue of separate waste collection, as no formal training was provided to sanitary workers on how to handle the waste during the pandemic. Arranging regular operator training is another difficult issue.
- Inconsistency in the supply of PPEs and disinfectants increases the likelihood of collection staff infection.
- Inadequate donning and doffing of PPE among waste workers may increase the spread among them.
- Failure to maintain social distancing measures at collection and treatment centers.
- Workers at the wastewater treatment plant may become infected if wastewater from healthcare facilities contains viruses. There is no door-to-door collection of staff from residential areas, and one-point collection is predominant, increasing the risk of infection.
- The use of separate vehicles for infected and noninfected waste transport, as well as a lack of disinfectant to clean the vehicles, increases the risk of virus spread.
- In towns and villages, there is a lack of proper monitoring, review, and verification systems.
- Challenges in implementing strategies and providing advice to people who handle waste on the ground.
- Difficulty in making the public aware of the importance and rules of waste segregation in a short period of time.
- Manual loading or unloading of waste is required in a few places, which increases the likelihood of contamination.

2.6 Effective waste management strategies

According to the WHO, the pandemic will last for a long time and will have a long-term impact on environmental sustainability. As a result, the entire world is focusing on prevention strategies and vaccine research. For the COVID-19 outbreak, most countries had become accustomed to or restructured their waste management strategies. In this regard, China can provide valuable inputs based on its experience in developing an all-inclusive waste disposal system that combines central disposal with on-site emergency waste disposal (such as mobile treatment and industrial kilns). The pandemic not only changed people's lifestyles, but it also prompted more thought about the human-nature relationship. Furthermore, there are a few critical issues regarding whether BMW management's existing systems can quickly restore waste processing, as well as the general population's ability to focus on waste reduction. These issues must be discussed and researched to improve the future. There is no single solution available for various countries, which differ in terms of infested cases and health resources, so various solutions should be implemented. There are numerous issues to be addressed when dealing with appropriate and actual waste management because it involves many departments and individuals, necessitating interdepartmental collaboration. To determine whether a country can handle a significant increase in BMW production, it should examine its existing strategies, rules, and structures, as well as the status of waste operations. The main issue is not how to frame the strategies and policies, but how to implement them successfully. To create a sustainable waste management system throughout and after the pandemic at the national or local levels, both policymakers and experts must consider at least seven thematic significance areas as identified by UNICEF (WHO & UNICEF, 2020).

2.6.1 Policy, regulatory, and guidelines

Even though international principles and strategies for managing BMW are available and widely referred to and followed by countries, this only provides a reasonable foundation in healthcare waste management during the COVID-19 pandemic. Supplementary rules and guiding principles will always be required to control waste generated from both healthcare and nonhealthcare facilities, especially when there is an increase in the generation of infected waste. During an emergency or pandemic, a solid plan with clearly defined roles and responsibilities for waste management is required, including an exigency plan and readiness for further improvement through a review of existing policies. It is critical to conduct a quick valuation of the countries or cities' existing facility systems to identify gaps and how to maximize the utilization of available facilities.

2.6.2 Handling of infectious waste

It is critical to understand the massive amount of healthcare waste produced during the pandemic, but scientific understandings, such as the virus's viability, should also be considered to create a sound waste management system (WHO & UNICEF, 2020). It must be ensured that potentially contaminated wastes are segregated, packaged, and stored properly. The frequency of collection should be determined by priority and waste type. Proper use of PPE, hand hygiene, and precautionary practices to protect waste employees' health must be maintained.

2.6.3 Suitable disposal methods

The selection of appropriate treatment options is relative and necessitates the investigation of numerous factors. It is not only limited to national and international principles, but also includes environmental and occupational safety, waste characteristics, appropriate technology costs, maintenance factors, and other related aspects. Many facilities in developing countries lack the most desirable waste treatment technologies. It is recommended that such countries have at least temporary or interim treatment methods in place, such as single-chamber incineration or automated pressure pulsing gravity autoclaves. In the event of an emergency pandemic, developing countries can implement these interim treatment options in accordance with national and international guiding principles. As a sustainable development concept coprocessing, open burning, and temporary storage as a momentary measure may be considered [Asian Development Bank (ADB), 2020].

2.6.4 Information, education, and communication

Healthcare workers are trained to handle healthcare waste, but a capacity building and awareness program for households and the public is now essential. A program that uses a user-friendly communication channel, such as radio, television, or websites with everyday displays, can raise public awareness about safe waste disposal.

2.6.5 Data management and learning

The first step in planning and policy development is data collection. As a result, the value of data is enormous. However, data availability is frequently difficult, particularly during an emergency when resources are allocated to immediate responses. The assistance of academic institutions in data collection is a very useful option. Regular discussion is a good practice that is required to exchange perspectives and lessons learned for continuous improvement during any emergency, particularly in pandemic situations.

2.6.6 Monitoring of segregation

The first and most important step in biomedical waste management, segregation, should be monitored. Segregation reduces waste generation as well as infection spread. As discussed by Mondal et al. (2015) (Mondal & Satyanarayana, 2018; Ramkrishna & Satyanarayana, 2017), the biomedical waste segregation deficiency index (BMWSD index) may be used to monitor segregation, which could be used for training needs, corrective and preventive action, and the study shows how the BMWSD index improved over the months.

2.6.7 Basic principles for managing waste

During an infectious disease pandemic, some basic ideologies and guidelines must be followed to manage infectious waste. Waste produced by healthcare facilities (BMW): These types of waste should be handled in accordance with national and international regulations. Enhanced treatment and disposal capacities should be identified in epidemic areas based on an analysis of waste capacity gaps, options available, modes of treatment and disposal, local constraints, and infection concerns for human health and environmental impacts. Household waste: Priority should be given to the continuity of existing waste management services. Contingency planning entails ensuring the availability of required human resources and equipment. Special care must be taken to reduce the risks to those performing waste operations. Increase awareness among waste management service providers to ensure an adequate supply of materials, as well as staff training and capacity building.

2.6.8 Fundraising and national and international collaboration

The local government should recognize the importance of fundraising activities in relation to BMW management. In the future, sufficient funds should be maintained to deal with such pandemic situations. Collaboration at the national and international levels with various agencies is required for the activities. During a pandemic, the author proposed a new waste management model. First, the capacity of both waste generators and waste handlers should be thoroughly assessed or planned. These will assist us in properly planning. The first two of the three R's (reduce, reuse, and recycle) may not be applicable during the COVID-19 pandemic, but recycling must be tried with appropriate technology. The main issue was a lack of incinerators. Suggestions for new incinerators should be considered to build large hospitals that are self-sufficient, with the ability to add capacity as needed. However, space constraints should be considered and, if possible, relaxed. More staff training is needed to handle waste while taking all possible precautions. During a pandemic, mobile and industrial incinerators may be used.

2.7 Conclusion and future directions

COVID-19 had a significant impact on economic development and health outcomes. Aside from the economic slowdown, COVID-19 has had a significant impact on daily wage and migrant workers (Chinmay et al., 2020; Sukanya & Chinmay, 2021). Waste management is now regarded as an essential public service, and proper waste management from households and medical facilities requires immediate attention. Contaminated waste can contaminate waste handlers because they are directly exposed to waste and use inadequate safety equipment, which can lead to a gradual increase in virus spread. Appropriate waste management also aids in the recycling of waste or the conversion of waste into a valuable product, such as energy, which will add value to economies aiming for sustainable development. The most significant challenges with COVID-19 management are its safe disposal from the point of generation to treatment facilities. There is great concern about the management of COVID-19 waste because positive cases are increasing rapidly, and hospitals are overburdened by such BMW, which could result in contamination among sanitary workers if precautions are not taken. Proper and systematic sanitization practices, as well as adequate PPE, should be required for workers involved in the handling and collection of BMWs. Staff should be properly and regularly trained to handle waste and prevent virus transmission through waste. Proper waste segregation in healthcare is required for any healthcare organization and should be strictly enforced. In these cases, monitoring methods based on the BMW segregation deficiency index (BMWSD index) may be used to assess training needs (Chinmay et al., 2020; Mondal & Satyanarayana, 2018; Ramkrishna & Satyanarayana, 2017; Sukanya & Chinmay, 2021). Public awareness for waste segregation at the point of generation in homes, before it can harm other masses, should be promoted, which will aid in the segregation of waste used in homes. Because everyone uses PPE that is made of plastic, future research should focus on biodegradable or environmentally friendly protective items. Regular monitoring is required when removing COVID-19 waste to prevent workers from becoming infected (Chinmay et al., 2020; Mondal et al., 2015; Mondal & Chakrabarty, 2022; Sukanya & Chinmay, 2021). To ensure the safety of municipal workers and rag pickers, communities must manage their waste properly while under quarantine. Municipal workers involved in handling the outbreak must have proper safety kits and training. There are numerous challenges, but appropriate strategies and careful handling can significantly reduce the risks.

References

ADEPT. (2020). *COVID-19–waste survey results, week commencing 18th May*. ADEPT www.adeptnet.org.uk/documents/covid-19-waste-survey-results-wc-18th-may.

Asian Development Bank (ADB). (2020). *Managing infectious medical waste during the COVID-19 pandemic*. Asian Development Bank (ADB) www.adb.org/publications/managing-medical-wastecovid1.

Association of Cities and Regions for Sustainable Resource Management. (2020). *Municipal waste management and COVID-19*. Association of Cities and Regions for Sustainable Resource Management.

Bashir, M. F., Ma, B., Bilal, K. B., Bashir, M. A., Tan, D., & Bashir, M. (2020a). Correlation between climate indicators and COVID-19 pandemic in New York, USA. *Science of the Total Environment, 728*, 138835. doi:10.1016/j.scitotenv.2020.138835.

Bashir, M. F., Ma, B., & Shahzad, L. (2020b). A brief review of socio-economic and environmental impact of COVID-19. *Air Quality, Atmosphere and Health, 1*, 1–7. doi:10.1007/s11869-020-00894-8.

Bogoch, Watts, A., Thomas-Bachli, A., Huber, C., Kraemer, M. U., & Khan, K. (2020). Pneumonia of unknown etiology in Wuhan, China: Potential for international spread via commercial air travel. *Journal of Travel Medicine, 27*. doi:10.1093/jtm/taaa008.

Brizi, A., & Biraglia, A. (2021). "Do I have enough food?" How need for cognitive closure and gender impact stockpiling and food waste during the COVID-19 pandemic: A cross-national study in India and the United States of America. *Personality and Individual Differences, 168*, 110396.

Calma, J. (2020). The COVID-19 pandemic is generating tons of medical waste. *The COVID-19 Pandemic is Generating Tons of Medical Waste*. https://www.theverge.com/2020/3/26/21194647/the-covid-19-pandemic-is-generating-tons-of-medical-waste.

Chakraborty, I., & Maity, P. (2020). COVID-19 outbreak: Migration, effects on society, global environment and prevention. *Science of the Total Environment, 728*, 138882. doi:10.1016/j.scitotenv.2020.138882.

Chinmay, C., Roy, S., Sharma, S., & Tran, T. A. (2020). *Environmental sustainability for green societies: The impact of the COVID-19 pandemic*. Springer Nature.

Clement J. Coronavirus global online traffic impact as of September 2020. 2020. www.statista.com/statistics/1105495/coronavirus-traffic-impact

Collivignarelli, M. C., Abbà, A., Bertanza, G., Pedrazzani, R., Ricciardi, P., & Miino, M. C. (2020). Lockdown for COVID-2019 in Milan: What are the effects on air quality? *Science of the Total Environment, 732*, 139280. doi:10.1016/j.scitotenv.2020.139280.

Conticini, E., Frediani, B., & Caro, D. (2020). Can atmospheric pollution be considered a co-factor in extremely high level of SARS-CoV-2 lethality in northern Italy? *Environmental Pollution*, 114465. doi:10.1016/j.envpol.2020.114465.

Feber, D., Lingqvist, O., & Nordigården, D. (2020). How the packaging industry can navigate the coronavirus pandemic. *McKinsey and Company*, 2–3. www.mckinsey.com/industries/paper-forest-productsand-packaging/our-insights/how-the-packaging-industry-can-n.

Espejo, W., Celis, J. E., Chiang, G., & Bahamonde, P. (2020). Environment and COVID-19: Pollutants, impacts, dissemination, management and recommendations for facing future epidemic threats. *Science of the Total Environment, 747*, 141314. doi:10.1016/j.scitotenv.

European Commission. (2020). *Waste management in the context of the coronavirus crisis*. European Commission.

Gautam, S. (2020). COVID-19: Air pollution remains low as people stay at home. *Air Quality, Atmosphere and Health, 13*, 853–857. doi:10.1007/s11869-020-00842-6.

Goodell, J. W., & Huynh, T. L. D. (2020). Did congress trade ahead? considering the reaction of U.S. industries to COVID-19. *Finance Research Letters*, 101578. doi:10.1016/j.frl.2020.101578.

Grodzińska-Jurczak, M., Krawczyk, A., Jurczak, A., Strzelecka, M., Rechciński, M., & Boćkowski, M. (2020). Environmental choices vs. COVID-19 pandemic fear – plastic governance re-assessment. *Socialist Register, 4*(2), 49–66.

Hellewell, J., Abbott, S., Gimma, A., Bosse, N. I., Jarvis, C. I., Russell, T. W., … Eggo, R. M. (2020). Feasibility of controlling COVID-19 outbreaks by isolation of cases and contacts. *The Lancet Global Health, 8*, e488–e496. https://doi.org/10.1016/S2214-109X(20)30074-7.

Hunter. Food study special report. 2020. www.slideshare.net/HUNTERNY/hunter-food-study-special-report-america-gets-cooking-231713331.

Ikiz, E., Maclaren, V. W., Alfred, E., & Sivanesan, S. (2021). Impact of COVID-19 on household waste flows, diversion and reuse: The case of multi residential buildings in Toronto. *Canada. Resources, Conservation and Recycling, 164*, 105111.

Ilyas, S., Srivastava, R. R., & Kim, H. (2020). Disinfection technology and strategies for COVID-19 hospital and bio-medical waste management. *Science of the Total Environment, 749*, 141652.

Jribi, S., Ben Ismail, H., Doggui, D., & Debbabi, H. (2020). COVID-19 virus outbreak lockdown: What impacts on household food wastage? *Environment, Development and Sustainability, 22*(5), 3939–3955.

Klemeš, J. J., Fan, Y. V., Tan, R. R., & Jiang, P. (2020). Minimising the present and future plastic waste, energy and environmental footprints related to COVID-19. *Renewable and Sustainable Energy Reviews, 127*, 109883. doi:10.1016/j.rser.202.

Knowland, K.E., Keller, C.A., Wales, P.A., Wargan, K., Coy, L., Johnson, M.S., ... Pawson, S. (2022). NASA GEOS Composition Forecast Modeling System GEOS-CF v1.0: Stratospheric Composition. Journal of advances in modeling earth systems, 14(6), e2021MS002852. https://doi.org/10.1029/2021MS002852.

Lal, P., Kumar, A., Kumar, S., Kumari, S., Saikia, P., Dayanandan, A., & AdhikariD, K. M. L. (2020). The dark cloud with a silver lining: Assessing the impact of the SARS COVID-19 pandemic on the global environment. *Science of the Total Environment, 732*, 139297.

Luan, P. T., & Ching, C. T. S. (2020). A reusable mask for coronavirus disease 2019 (COVID-19). *Archives of Medical Research, 51*, 455–457. doi:10.1016/j.arcmed.2020.04.001.

Messner, W. (2020). The institutional and cultural context of cross-national variation in COVID-19 outbreaks. *MedRxiv*. doi:10.1101/2020.03.30.20047589.

Mihai, F. C. (2020). Assessment of COVID-19 waste flows during the emergency state in Romania and related public health and environmental concerns. *International Journal of Environmental Research and Public Health, 17*(15), 5439.

Mizumoto, K., Kagaya, K., Zarebski, A., & Chowell, G. (2020). Estimating the asymptomatic proportion of coronavirus disease 2019 (COVID-19) cases on board the diamond princess cruise ship. *Eurosurveillance, 25*(10), 1–5. doi:10.2807/15607917.es.2020.25.10.20.

Mondal, R., Chakrabarty, C., Effect of COVID-19 pandemic on waste management system and infection control' (Healthcare Technologies, 2022), 'Smart health technologies for the COVID-19 pandemic: Internet of medical things perspectives, IET Digital Library, pp. 385–403, doi: 10.1049/PBHE042E_ch14, https://digital-library.theiet.org/content/books/10.1049/pbhe042e_ch14

Mondal, R., Sarkar, N., Satyanarayana, P., & Lingaiah, A. (2015). Work study of bio-medical waste transportation from point of generation to storage in a tertiary care hospital. *Asian Journal of Management, 6*(4), 331–335. doi:10.5958/2321-5763.2015.00049.9.

Mondal, R., & Satyanarayana, P. (2018). *Bio-medical waste segregation monitoring in healthcare establishment – An Indian scenario*. LAP LAMBERT Academic Publishing.

Nemudryi, A., Nemudraia, A., Surya, K., Wiegand, T., Buyukyoruk, M., & Wilkinson, R., et al. (2020). Temporal detection and phylogenetic assessment of SARS-CoV-2 in municipal wastewater. Cell Reports Medicine, 1(6), 10098. doi:10.1016/j.xcrm.2020.100098. Nemudryi, A., Nemudraia, A., Wiegand, T., Surya, K., Buyukyoruk, M., Cicha, C., Vanderwood, K.K., Wilkinson, R., & Wiedenheft, B. (2020). Temporal Detection and Phylogenetic Assessment of SARS-CoV-2 in Municipal Wastewater. *Cell Reports Medicine, 1*(6), 100098. https://doi.org/10.1016/j.xcrm.2020.100098.

Nishiura, H., Kobayashi, T., Suzuki, A., Jung, S.-M., Hayashi, K., & Kinoshita, R., et al. (2020). Estimation of the asymptomatic ratio of novel coronavirus infections (COVID-19). International Journal of Infectious Diseases, 94, 154–155. doi:10.1016/j.ijid.20.Nishiura, H., Kobayashi, T., Miyama, T., Suzuki, A., Jung, S.M., Hayashi, K., Kinoshita, R., Yang, Y., Yuan, B., Akhmetzhanov, A.R., & Linton, N.M. (2020). Estimation of the asymptomatic ratio of novel coronavirus infections (COVID-19). *International journal of infectious diseases : IJID : official publication of the International Society for Infectious Diseases, 94*, 154–155. https://doi.org/10.1016/j.ijid.2020.03.020.

Naughton, C. (2020). Will the COVID-19 pandemic change waste generation and composition? The need for more real-time waste management data and systems thinking. *Resources, Conservation and Recycling, 162*, 105050.

O'Bannon, D., Xagoraraki, I., & O'Brien, E. (2020). Wastewater-based epidemiology for early detection of viral outbreaks. *Women in water quality* (pp. 75–79). Springer Nature Switzerland. doi:10.1007/978-3-030-17819-2.

Patrício Silva, A. L., Prata, J. C., Walker, T. R., Duarte, A. C., Ouyang, W., Barcelo, D., & Rocha-Santos, T. (2021). Increased plastic pollution due to COVID-19 pandemic: Challenges and recommendations. *Chemical Engineering Journal, 405*, 126683.

Peng, J., Wu, X., Wang, R., Li, C., Zhang, Q., & Wei, D. (2020). Medical waste management practice during the 2019–2020 novel coronavirus pandemic: Experience in a general hospital. *American Journal of Infection Control, 48*(8), 918–921.

Poon, L. L. M., & Peiris, M. (2020). Emergence of a novel human coronavirus threatening human health. *Nature Medicine, 26*(3), 317–319.

Prata, J. C., Silva, A. L. P., Walker, T. R., Duarte, A. C., & Rocha-Santos, T. (2020). COVID-19 pandemic repercussions on the use and management of plastics. *Environmental Science & Technology, 54*(13), 7760–7765.

Ragazzi, M., Rada, E. C., & Schiavon, M. (2020). Municipal solid waste management during the SARS-COV-2 outbreak and lockdown ease: Lessons from Italy. *Science of the Total Environment, 745*, 141159.

Rahman, M. M., Bodrud-Doza, M., Griffiths, M. D., & Mamun, M. A. (2020). Biomedical waste amid COVID-19: Perspectives from Bangladesh. *The Lancet Global Health, 8*(10), e1262. https://www.thelancet.com/journals/langlo/article/PIIS2214-109X(20)30349-1/abstract.

Ramkrishna, M., & Satyanarayana, P. (2017). A study on bio-medical waste segregation monitoring in a tertiary care hospital at Andhra Pradesh. *Asian Journal of Management, 8*(1), 73–77. doi:10.5958/2321-5763.2017.000117.

Ramteke, S., & Sahu, B. L. (2020). Novel coronavirus disease 2019 (COVID- 19) pandemic: Considerations for the biomedical waste sector in India. *Case Studies in Chemical and Environmental Engineering*, 100029. doi:10.1016/j.cscee.2020.100029.

Rodríguez-Pardo, J., Fuentes, B., Alonso de Leciñana, M., Campollo, J., Calleja Castaño, P., & Carneado Ruiz, J., et al. (2020). Acute stroke care during the COVID-19 pandemic. Ictus Madrid Program recommendations. Neurologia, 35, 258–263.Rodríguez-Pardo, J., Fuentes, B., Alonso de Leciñana, M., Campollo, J., Calleja Castaño, P., Carneado Ruiz, J., Egido Herrero, J., García Leal, R., Gil Núñez, A., Gómez Cerezo, J.F., Martín Martínez, A., Masjuán Vallejo, J., Palomino Aguado, B., Riera López, N., Simón de Las Heras, R., Vivancos Mora, J., Díez Tejedor, E., & en nombre del Grupo Multidisciplinar del Plan Ictus Madrid (2020). Acute stroke care during the COVID-19 pandemic. Ictus Madrid Program recommendations. Atención al ictus agudo durante la pandemia por COVID-19. Recomendaciones Plan Ictus Madrid. *Neurologia, 35*(4), 258–263. https://doi.org/10.1016/j.nrl.2020.04.008.

Rupani, P. F., Nilashi, M., Abumalloh, R. A., Asadi, S., Samad, S., & Wang, S. (2020). Coronavirus pandemic (COVID-19) and its natural environmental impacts. *International Journal of Environmental Science and Technology, 17*, 4655–4666.

Sarkodie, S. A., & Owusu, P. A. (2020a). Global assessment of environment,health and economic impact of the novel coronavirus (COVID-19). *Environment, Development and Sustainability*, 1–11.

Sharif, A., Aloui, C., & Yarovaya, L. (2020). COVID-19 pandemic, oil prices, stock market, geopolitical risk and policy uncertainty nexus in the U.S. economy: Fresh evidence from the wavelet-based approach. *International Review of Financial Analysis, 70*, 101496. doi:10.1016/j.irfa.2020.101496.

Singh, N., Tang, Y., Zhang, Z., & Zheng, C. (2020). COVID-19 waste management: Effective and successful measures in Wuhan, China. *Resources, Conservation and Recycling, 163*(105071), 1–3. doi:10.1016/j.resconrec.2020.105071.

Somani, M., Srivastava, A. N., Gummadivalli, S. K., & Sharma, A. (2020). Indirect implications of COVID-19 towards sustainable environment: An investigation in Indian context. *Bioresource Technology Reports, 11*, 100491.

Sukanya, R., & Chinmay, C. (2021). Panic buying situation during COVID-19 global pandemic. *Journal of Information Technology Management, 13*(2), 231–244. doi:10.22059/jitm.2021.80625.

Thakur, V. (2021). Framework for PESTEL dimensions of sustainable healthcare waste management: Learnings from COVID-19 outbreak. *Journal of Cleaner Production, 287*.

van Doremalen, N., et al. (2020). Aerosol and surface stability of SARS-CoV- 2 as compared with SARS-CoV-1. New England Journal of Medicine, 382(16), 1564–1567.van Doremalen, N., Bushmaker, T., Morris, D.H., Holbrook, M.G., Gamble, A., Williamson, B.N., Tamin, A., Harcourt, J.L., Thornburg, N.J., Gerber, S.I., Lloyd-Smith, J.O., de Wit, E., & Munster, V.J. (2020). Aerosol and Surface Stability of SARS-CoV-2 as Compared with SARS-CoV-1. *New England Journal of Medicine, 382*(16), 1564–1567. https://doi.org/10.1056/nejmc2004973.

UK EnvironmentAgency. (2020). *COVID-19 and storing waste at unpermitted sites due to exceeding your storage limits: RPS C17*. UK Environment Agency.

United Nations. (2020). *Sustainable development goal 11: Make cities and human settlements inclusive, safe, resilient, and sustainable*. United Nations.

Vanapalli, K. R., Sharma, H. B., Ranjan, V. P., Samal, B., Bhattacharya, J., Dubey, B. K., & Goel, S. (2021). Challenges and strategies for effective plastic waste management during and post COVID-19 pandemic. *Science of the Total Environment, 750*, 141514.

Water, sanitation, hygiene and waste management for the COVID-19 virus. Interim guidance, WHO/2019- n CoV/IPC_WASH/2020.2 (pp. 1–6). (2020).

Wang, D., Hu, B., Hu, C., Zhu, F., Liu, X., Zhang, J., … Xiong, Y. (2020). Clinical characteristics of 138 hospitalized patients with 2019 novel coronavirus–infected pneumonia in Wuhan, China. Jama, 323(11), 1061–1069. doi:10.1001/jama.Wang, D., Hu, B., Hu, C., Zhu, F., Liu, X., Zhang, J., Wang, B., Xiang, H., Cheng, Z., Xiong, Y., Zhao, Y., Li, Y., Wang, X., & Peng, Z. (2020). Clinical Characteristics of 138 Hospitalized Patients With 2019 Novel Coronavirus-Infected Pneumonia in Wuhan, China. *JAMA, 323*(11), 1061–1069. https://doi.org/10.1001/jama.2020.1585.

World Health Organization. (2016). *Safe healthcare waste management, policy paper, department of protection of the human environment water, sanitation and health*. World Health Organization.

Yang, L., Yu, X., Wu, X., Wang, J., Yan, X., Jiang, S., & Chen, Z. (2021). Emergency response to the explosive growth of healthcare wastes during COVID-19 pandemic in Wuhan. *China. Resources, Conservation and Recycling, 164*, 105074.

Yu, H., Sun, X., Solvang, W. D., & Zhao, X. (2020). Reverse logistics network design for effective management of medical waste in epidemic outbreaks: Insights from the coronavirus disease 2019 (COVID-19) outbreak in Wuhan (China). *International Journal of Environmental Research and Public Health, 17*.

Zambrano-Monserrate, M. A., & Ruano, M. A. (2020). Has air quality improved in Ecuador during the COVID-19 pandemic? A parametric analysis. *Air Quality, Atmosphere and Health, 13*, 929–938. doi:10.1007/s11869-020-00866-y.

Zand, A.D., & Heir, A.V. (2020). Emerging challenges in urban waste management in Tehran, Iran during the COVID-19 pandemic. Resources, Conservation and Recycling, 162, 105051.Zand, A.D., & Heir, A.V. (2020). Emerging challenges in urban waste management in Tehran, Iran during the COVID-19 pandemic. *Resources, conservation, and recycling, 162*, 105051. https://doi.org/10.1016/j.resconrec.2020.105051.

CHAPTER

Current trend of biomedical waste generation in pre- and post-COVID-19 pandemic

3

Alok Raghav[a], Goo Bo-Jeong[a], Poonam Khanna[b], Manish Singh[c] and Saurabh Mishra[d]

[a] *Department of Anatomy and Cell Biology, Lee Gill Ya Cancer and Diabetes Institute, Gachon University, Incheon, South Korea,* [b] *School of Public Health, Post Graduate Institute of Medical Education and Research, Chandigarh, Punjab, India,* [c] *Department of Neurosurgery, GSVM Medical College, Kanpur, Uttar Pradesh, India,* [d] *Department of Biochemistry, Chatrapati Shahu Ji Maharaj University, Kanpur, Uttar Pradesh, India*

3.1 Introduction

Coronaviruses, which are members of the Coronaviridae family, are responsible for respiratory infections in birds and animals, including bats, camels, and masked palm civets (Fehr & Perlman, 2015). Broad host animals may exhibit diverse coronavirus symptoms of infection and tissue tropism (Chen et al., 2020). Three significant coronavirus waves have been identified and documented to date, with the latest one being caused by the 2019 novel coronavirus (2019-nCoV, also known as SARS-CoV-2 lately), which is responsible for causing COVID-19 (The New York Times, 2020). The World Health Organization (WHO) confirmed a worldwide health emergency in response to the new coronavirus epidemic in Wuhan, a city in China's Hubei province, on January 30, 2020. SARS-CoV-2 has the potential to spread globally and spark a pandemic outbreak, according to the WHO, which acknowledged this on February 24, 2020. The WHO subsequently announced COVID-19 a pandemic on March 11, 2020 (Vaughan, 2020; The New York Times, 2020). COVID-19, caused by SARS-CoV-2, is one of a vast number of extremely varied, encapsulated, positive-sense, and single-stranded RNA viruses (Ilyas et al., 2020). Coronaviruses have one of the biggest RNA viral genomes, ranging in size from around 26 to 32 kilobases (Agamuthu & Barasarathi, 2021). Full-genome sequencing and phylogenetic studies revealed that the coronavirus responsible for COVID-19 is a betacoronavirus in the identical subgenus as the severe acute respiratory syndrome (SARS) virus (as well as multiple bat coronaviruses), but in an entirely different clade (Zhou et al., 2020). It looks possible that bats are a major source of the virus because two bat viruses share an identical RNA sequence. It is unclear if the COVID-19 virus spreads through infected bats or via another route (for example, through an intermediary host). This virus has similarities with the two earlier viruses that caused the Middle East respiratory

Biomedical Waste Management in the COVID-19 Era: A Global Scenario.
DOI: https://doi.org/10.1016/B978-0-443-21841-5.00001-1

syndrome (MERS) epidemic in 2012 and the SARS outbreak in 2003 (Capoor & Parida, 2021). While symptoms often appear 5 to 6 days after viral infection, they can appear up to 14 days later (Agamuthu & Barasarathi, 2021). The WHO reported 5,020,204 people dying as of November 4, 2021, out of 247, 968, 227 confirmed cases of COVID-19. The ecology and the climate have been impacted by COVID-19 in both positive and negative ways (Ngoc et al., 2022). According to statistics on the effects of COVID-19 on the world as of the online reporting on May 23 2021, which includes a total of 16,70,60,227 confirmed cases of COVID-19 with 34,69,014 deaths in a total of 222 nations of the world, the United States, India, Brazil, France, Turkey, Russia, the United Kingdom, Italy, Germany, and Spain are the top ten most affected countries. The continental bifurcation of COVID-19 demonstrates the total number of confirmed cases in Europe as 4,61,62,211 (deaths 10,60,272); in North America as 3,94,61,755 total cases (8,84,707 deaths); in Asia as 4,89,29,156 total cases (6,42,543 deaths); in South America as 2,76,47,386 total cases (7,51,694 deaths); in Africa as 47,91,745 total cases (1,28,540 deaths); and Oceania with 67,253 total cases (1243 deaths) (Worldometer, n.d.).

Any sustainable healthcare industry must have proper management of biomedical waste (BMW). With the COVID-19 pandemic outbreak, which left every hospital and medical facility overrun with patients, managing the enormous volume of produced BMW effectively became a global challenge. It is widely known from previous epidemics and breakouts that the volume of BMW produced during every health crisis is far more than the typically produced medical waste (Khosla et al., 2022). Any waste produced when diagnosing, treating, or immunizing people or animals, doing studies on them, or while producing or evaluating medical products is referred to as BMW. A growing nation like India produces 1.5 to 2 kg of BMW per bed every day. After the beginning of COVID-19, there was a dramatic increase in the amount of COVID-19 BMW produced from designated COVID-19 sites, reaching 7.76 kg/COVID bed/day. Lockdown caused elective admissions in non-COVID wards to cease; as a result, waste generation from COVID-designated locations has been reported (Gowda et al., 2022). According to data released by the Central Pollution Control Board (CPCB) in 2018, India created 517 tons of BMW per day in 2016 and around 501 tons per day in 2015, with about 4%–5% of that quantity being untreated (Khosla et al., 2022; Pillay & Barnes, 2020). According to the CPCB's annual statement for 2018/2019, 557 tons of BMW were produced daily in 2017; 517 of those tons were processed. There are 238,170 healthcare facilities in the nation; 87,267 of them have beds, while the remaining 151,208 are nonbedded HCFs that produce BMW. There are 198 common BMW disposal facilities (CBMWFs) that have been authorized nationwide, and another 28 are currently being built (Chand et al., 2021). In South Asian nations, there is a huge discrepancy between urban and rural BMW management. The majority of healthcare institutions do not adhere to BMW management standards. A significant number of healthcare facilities, including 82% of primary, 60% of secondary, and 54% of tertiary health centers, did not have a reliable BMW management system, according to the INCLEN Program Evaluation Network (IPEN) Study Group,

New Delhi. In remote places, clinics are administered by unlicensed medical professionals who lack the hands-on abilities required to operate a BMW and who also struggle to understand color coding. Additionally, there were issues with the storage of garbage, the maintenance of records, and the sorting of waste into distinct sorts (Shammi et al., 2021). BMW is all the waste generated by healthcare facilities, medical laboratories, and biomedical research facilities, as well as waste from minor or scattered sources. Although hospitals produce the bulk of healthcare waste by volume, they are a small fraction of the total number of sources.

The rapid spread of the COVID-19 pandemic has resulted in a significant number of deaths, which has an adverse effect on mental health and raises societal issues because of governmental limitations (such as imprisonment and curfews) (Miyah et al., 2022). Therefore, the time frame of the sanitary measures implemented to control the virus may be a contributing factor in the symptoms of anguish, sadness, anxiety, depression, posttraumatic stress disorder, frustration, and suicide. It is advised to train psychiatrists and social service professionals in dealing with the consequences of outbreaks and health disasters and to sensitize patients to visit psychiatrists to lower the risk of contagion to overcome these mental health problems (Cullen et al., 2020; Pillay & Barnes, 2020). Black, Asian, and minority ethnic communities have suffered unfairly in the UK and the USA. The persistent use of personal protective equipment (PPE) is crucial to reducing contact at home transmission, according to those dealing with other respiratory illnesses. Whereas it is the primary objective to deal with the requirements of the front-line's healthcare professionals during the COVID-19 epidemic (CDC COVID-19 Response Team 2020; Lo & De Angelis, 2020). Depending on the conditions, threat, or extent of the pandemic, government officials should suggest or implement steps on a national and subnational level. They should also periodically review the scenario as it changes. Communities may also modify policies as necessary, taking into account the local culture, housing conditions, and accessibility to facilities and amenities (World Health Organization, 2020). Although this may be acceptable in the early stages of an epidemic when health services prioritize testing, stopping the spread of the disease, and providing crucial patient care, psychological and psychiatric requirements should not be ignored at any stage of dealing with a pandemic (Cullen et al., 2020). Second, beyond health professionals, choices concerning who is tested or that take treatment have to be made with the goal of preventing SARS-CoV-2 transmission (public health), safeguarding those who are most vulnerable, addressing societal needs, and advancing social justice. Prioritizing resources for those in restricted spaces (such as homeless shelters, prisons, and nursing homes), wherever the virus can pass quickly from person to person, may be necessary to protect the health of everyone. It might be necessary to focus resources on areas experiencing low (Gostin et al., 2020).

Improper treatment and disposal of healthcare waste poses serious hazards of secondary disease transmission due to exposures to infectious agents among waste pickers, waste workers, health workers, patients, and the community in general, where waste is improperly disposed of WHO (n.d.b).

3.2 Classification

The WHO has classified healthcare waste into the following types (WHO, n.d.a; WHO, n.d.b) (Table 3.1).

- *Infectious waste*: Wastes like blood and other bodily fluids, wasted lab samples, cultures, and other infectious agents from autopsies or diagnoses, animal wastes (from veterinary facilities), trash from sick patients like swabs, abandoned medical equipment, etc., are included in this category.
- *Chemical waste*: In addition to wasted X-ray films, elemental mercury, lead aprons, foils, scrap amalgam material, and other cleaning agents, chemical waste also includes disinfectants, laboratory chemicals, and solvents that have reached the end of their useful lives.
- *Cytotoxic waste*: It is a kind of waste that contains genotoxic particles, such as cytotoxic drugs, having teratogenic or carcinogenic impacts. It causes several health issues both within and outside the hospital environment. Antineoplastic or cytotoxic drugs prevent the growth of living cells when administered to the human body. Genotoxic waste covers these cytotoxic drugs as well as other radioactive materials.
- *Biotechnological or microbiological wastes*: These biological wastes consist of specimens of human or animal cell cultures, laboratory cultures, infectious particles arising out of biomedical research, toxins generated, and other equipment or devices used in the culture.
- *Pathological waste*: It includes anatomical wastes like body parts, human fetuses, animal carcasses, tissues, organs, fluids like blood, saliva, and extracted teeth, etc. Some of these pathogens are highly dangerous, possessing increased levels of pathogenicity and resistance to treatment.
- *Pharmaceutical waste*: Medicines such as ointments, pills, medications, syrups, and other liquid solutions like massage oils are among the wastes from pharmaceutical items that are thrown away, expired, or only partially utilized.
- *Pressurized containers*: Containers containing gas or pressurized liquids, containers for powdered materials, aerosol cans, and other gas-like containers fall under this category.
- *Radioactive waste*: This category includes wastes produced by radiation and research facilities, such as glassware, absorbent sheets, patient excreta and bodily fluids polluted with radionuclides, contaminated packaging, etc.
- *Sharps*: If not properly controlled, these compounds have a considerable danger of infecting healthcare workers and professionals with fatal infections and inflicting cuts or wounds. It includes infusion sets as well as needles, blades, scalpels, nails, scissors, razors, and knives, as well as shards of shattered glass and bone.
- *General or nonhazardous waste*: It is made up of waste that does not provide any physical, chemical, radiological, or biological risks.

  Other categories:
- *Soiled waste*: it consists of contaminated substances such as used cotton, bandages, plaster casts, dressings, lined beddings, etc.

Table 3.1 Biomedical waste categorization according to WHO and the European Union.

–	Category (examples)	World health organization (WHO)	EU	Source
Hazardous	Sharps	Sharps	Sharps	Hospitals, clinics, laboratories, blood banks, nursing homes, veterinary clinics, and labs
	Organic matter, including body parts and blood	Pathological	Human tissue, body parts, organs, and blood preserves and bags	Hospitals, clinics, laboratories, mortuaries and autopsy facilities, veterinary clinics, and labs
	Waste with restrictions in collection and disposal due to infectivity	Infectious	Human and animal infectious	Hospitals, clinics, and laboratories
	Waste with no restrictions or special requirements for collection and disposal due to infectivity (e.g., plasters, casts, dressings, bed sheets, disposable clothing, etc.)	Infectious	Infectious	Hospitals, clinics, and laboratories

(*continued on next page*)

Table 3.1 Biomedical waste categorization according to WHO and the European Union—cont'd

–	Category (examples)	World health organization (WHO)	EU	Source
	Dangerous chemical materials and substances	Chemical	Chemical	Hospitals, clinics, and laboratories
	Other chemicals	Chemical	Chemical/unused hazardous medicines	Hospitals, clinics, and laboratories
	Cytotoxic and cytostatic medicines	Cytotoxic	Discarded unused medicines	Hospitals and laboratories
Nonhazardous	Other chemicals (nonhazardous)	Pharmaceutical	Unused nonhazardous medicines	Hospitals, clinics, and laboratories
	Dental clinics (care centers) amalgam waste	Amalgam (tooth filling) waste from dental clinics/centers	Amalgam waste from dental clinics/centers	Dental care centers and clinics

Shammi et al. (2021), under Creative Commons Attribution License (CC BY).
Based on WHO and European Union biomedical waste categorization frameworks.
WHO, *World Health Organization.*

- *Wastes containing heavy metals*: It includes litters that consist of heavy metals such as mercury, lead, cadmium, etc., and their derivatives like discarded thermometers, manometers, and batteries.
- *Food wastes*: this category of waste consists of food particles that directly contact persons with infectious diseases.
- *Incinerated ash*: It contains ashes generated from the incineration process of BMW.

3.3 Current trend of biomedical waste generation in pre- and post-COVID-19

Worldwide, the production of medical waste is increasing quickly, which might pose a serious concern to both people and the environment. According to the WHO (2021), 85% of medical wastes are harmless, 10% are infectious (waste from sick patients, diagnostic tests, etc.), and 5% are dangerous (chemical and radioactive). A combination of its infectious, poisonous, or radioactive properties, the remaining 15% was deemed dangerous. Hazardous medical waste in particular could include potentially dangerous bacteria that might infect hospital patients, healthcare professionals, and members of the general public. Additionally, between 20% and 30% of medical waste may contain plastic trash, one of the flammable components (Tsai, 2021). Unfortunately, the coronavirus epidemic of 2019, also known as the COVID-19 pandemic, brought practically all aspects of society, systems, and life on the globe to a complete stop during the past few years. Due to the heavy consumption of disposable products and personal shielding safety equipment like PPE for both those with and without infection, as well as front-line workers, etc. During this time, a surge in the generation of a huge volume of diverse wastes at an unprecedented rate occurred (Luhar et al., 2022). However, the increased production of BMW has led to the creation of undue demand on the incinerator facilities, which has limited the widespread utilization of incineration. The amount of BMW in the regions of Hubei and Wuhan alone in China has grown by 370% and 600%, respectively, and now significantly surpasses the greatest incineration potential the nation has to offer (Dey et al., 2023). Nearly 9% of air pollution, 10% of smog formation, 10% of greenhouse gas emissions, 12% of acid rain leading to ozone depletion, and the discharge of cancer-causing and noncarcinogenic volatile hazardous chemicals worldwide are caused by healthcare organizations (Manzoor & Sharma, 2019).

3.4 Potential health risk associated with biomedical waste

Hazardous waste is anything that has the ability to affect humans in noninfectious ways. This can include things like chemicals (medical and industrial), old drugs, and sharps (needles, scalpels, and lancets, etc.). Medical waste that poses carcinogenic, pathogenic, and poisonous that risk of health. It may be divided into a few categories, including pharmaceutical waste, genotoxic waste, and chemical waste, all of which include plastic garbage (Khan et al., 2019). Syringes or vials containing cytostatic medications or substances that have been utilized for patient treatment are referred to as genotoxic and pathological wastes. Wastes having a high concentration of heavy metals, such as mercury, typically come from chemical and medicinal wastes. Additionally, the plastic garbage removed from landfills needs to be sterilized. This is because the surface of the plastic contains a lot of ash, pollutants, heavy metals, organic pollutants, and inorganic pollutants that might lead to biological reactions or chemical degradation (Acharya et al., 2014; Canopoli et al., 2018).

COVID-19 generated BMW or special bags for patients, including infected tissue face masks, gloves, and other materials. Plastics of many kinds, including polyethylene (PE), polypropylene (PP), polyethylene terephthalate (PET), polyvinyl chloride (PVC), and polystyrene (PS), are manufactured in numerous sectors. PE and PET are the two plastics that are most frequently encountered in plastic waste (PW) streams (Awoyera & Adesina, 2020). As everyone is aware, the volume of medical waste produced during the COVID-19 epidemic has a significant influence on the handling of plastic garbage. Regulating plastic garbage produced by residents, notably the unexpected increase in medical waste, has proven challenging in nations like China, India, Pakistan, and Bangladesh (Marinković et al., 2008). Infected people also overburdened medical facilities to the point where patients were recommended to self-isolate and quarantine at home, which exacerbated the problem of domestic toxic waste. Various public health protection guidelines and measures necessary to stop the spread of this global epidemic have collectively led to an increase in the demand for and the increased use of essential PPE, such as face masks, rubber boots, hand gloves, white gowns, hand sanitizers, etc. (Luhar et al., 2022). N-95 face masks, protective Tyvek suits, gloves, and face shields are frequently made of PP. However, these products allow the new coronavirus to persist for an extended length of time and release dioxin and other harmful substances into the air (Ramteke & Sahu, 2020). Since everyone is concerned about their health and cleanliness, the extensive use of single-use plastics (SUPs) during the COVID-19 epidemic has resulted in poor handling of worn PPE. In a single month, it was anticipated that 65 billion pairs of gloves and 129 billion face masks would be discarded worldwide (Prata et al., 2019).

The composition of polymers includes low-density polyethylene (LDPE), high-density polyethylene (HDPE), PET, PVC, and PP. Since mounting data suggest that people are exposed to plastics, long-term exposure to microplastics may add to the genetic instability burden of humans, which can lead to illnesses, most notably cancer. Despite the fact that our study was conducted in the laboratory and the chosen particles were able to reach bloodstream cells, our findings suggested that microplastics might be a significant danger factor for human health. More research is needed to understand the causes of DNA damage caused by such high microplastic concentrations and lower concentrations with longer exposure times (Çobanoğlu et al., 2021). Polycarbonate and PS are used to make PPE. The direct interaction of PP particles with cells can induce health issues by encouraging the generation of cytokines from immune cells rather than by direct toxicity to cells (Hwang et al., 2019). That leads to hormonal imbalance, cancer, nervous system diseases, and lowered immune levels. The main cause of the deadly internal damage in vertebrates and invertebrates is caused by PET microplastics. When this PET material burns, toxic chemicals such as nitric oxide, sulfur dioxide, and chlorofluorocarbons are released (Dhaka et al., 2022). The use of PPE may save lives during the period of the epidemic. Still, the collection, improper disposal, and littering of these PWs will lead the waste management systems to abruptly fail, severely polluting both the ecosystems of land and water. Moreover, during the COVID-19 epidemic, the flood of medical waste has increased each day. For instance, on the same day, February 24, 2020, the Chinese city of Wuhan generated 200 tons of medical waste (Marinković et al., 2008; Prata et al., 2019). Due to COVID-19, more polymer-made masks and other forms of PPE were created and utilized, which led to a pollutant is increased. In this regard, multiple studies have shown a variety of strategies to address the rise in waste produced by PPE kits and lessen their long-term effects on the environment and human health. Although a cotton mask can offer enough protection for most uses, the suggested N95 mask is constructed of polymers like PP and PET and is capable of filtering up to 95% of air particles with a size of 0.3 μm (Ray et al., 2022). During June and July 2020, PPE deliveries grew to 50.40 million (M) pieces (pcs) from 5.50 M the month before, having over 200 million pcs in store for urgent shipment to 138 nations, based on a WHO scenario report. Most of these goods are disposable and need to be thrown away. PPE is often used by the general population for protective apparel against viral infection, as well as by front-line healthcare workers for medical purposes. Since the outbreak of the pandemic, single-use plastic containers, gowns, and polyethene bags have become increasingly popular (Luhar et al., 2022; Siwal et al., 2021) (Fig. 3.1).

3.5 Potential management approaches for the dissemination of biomedical waste

Derisory infrastructure for BMW disposal is one of the major threats in developing nations, especially due to a lack of resources and awareness. During the pandemic, everyone gave priority to human health over other issues. However, if proper attention is not given, the threat of BMW poses a severe threat to human health, along with the environment, which will have adverse consequences. Though developing nations responded well to implementing the BMW management policies posed by WHO, the increased burden of BMW contributed to the ineffective implementation of such guidelines. The following management approaches should be implemented for effective dissemination of waste.

(I) The SUPs and BMW should be disposed of separately to inhibit the progression of the virus and to curtail the adverse threat to the environment and human health. Implementation of color-coded bins and bags should be ensured at the site of generation so that effective segregation can be done along with disposal (Datta et al., 2018; Das et al., 2021). The BMW management guidelines,

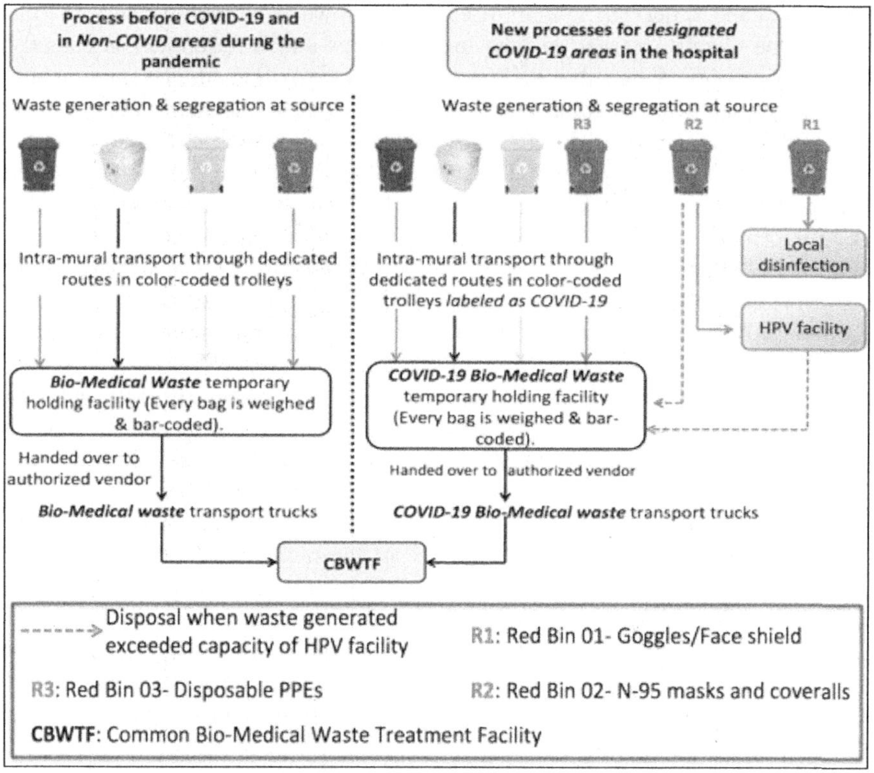

FIGURE 3.1

Prepandemic biomedical waste management process vis-à-vis process re-engineering for handling COVID-19 biomedical waste.

Data from Gowda et al. (2022), under Creative Commons Attribution License 4.0).

as described in previously published literature, should be implemented for adequate segregation, collection, storage, transportation, and disposal of the generated waste (CPCB 2020; Central Pollution Control Board CPCB, 2016). It is also advisable that, instead of using transitional or local policies, universal policies should be implemented homogenously. The local administrative authorities should provide the basic infrastructure, training, and awareness to the manpower at the generation site. Moreover, color-coded bins should also be placed at the community spots.

(II) Moreover, monetary investment in scaling up the infrastructure is needed to hire skilled professionals and to improve the disposal capacity, as suggested by the previously published literature (Deepak et al., 2021; Shammi et al., 2020). The implementation of mobile equipment with the possibility of expanding along

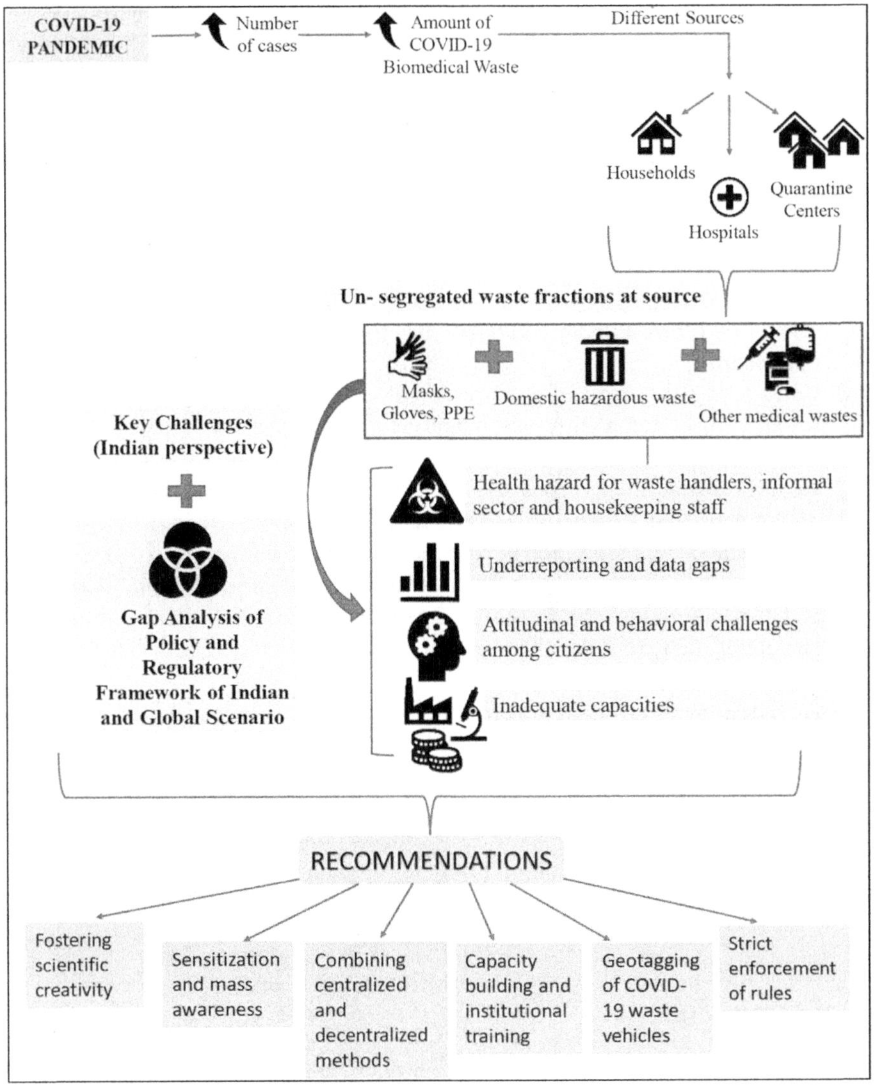

FIGURE 3.2

Management approaches of BMW.

Data from Khosla et al. (2022), under Creative Commons Attribution License (CC BY).

with capacity is the need of the current time to tackle the COVID-19 pandemic-like situation. During the pandemic, several mobile incinerators were installed in Wuhan, China, to overcome the problem of BMW.

(III) There is a need to increase the capacity of incinerators installed at common BMW facilities. Currently, most of the incinerators are modular type with low capacity; however, massive generation of BMW requires modular, rotatory, and fluidized bed incinerators depending upon the population size and quantity of BMW generated after the COVID-19 pandemic. It is also a matter of serious concern that all installed incinerators should be equipped with appropriate air pollution control and monitoring devices. Activated carbon filters and devices should be installed along with the incinerators to remove the hazardous dioxins, furans, and other persistent organic pollutants.

(IV) Considering the current approach for disposal of SUPs and PPEs, it is mandatory to increase screening of waste at aquatic, terrestrial, and aerial levels after the COVID-19 pandemic. Awareness should be spread regarding the hazardous effects of PPEs, plastic additives, and Persistent organic pollutants (POPs) degradation.

(V) Considering the need of the present time related to human health, there is a requirement for a fully automated and digital system to tackle the BMW disposal issues.

(VI) Health care facilities, especially tertiary care centers, must adopt decontamination approaches for recycling and reuse of some of the waste, including the PPEs. The decontamination approaches include washing, UV treatment, using hydrogen peroxide vapors, heat inactivation, gamma irradiation, alcohol use, and ethylene oxide (Ilyas et al., 2020; Kampf et al., 2020; Khoo et al., 2021; Liu et al., 2021; Mohanan et al., 2021; Rowan & Laffey, 2020). This approach significantly reduces the generation of BMW.

(VII) Advanced methods, including hydrothermal carbonization, can be a good alternative to incineration for the disposal of BMW (Bhar et al., 2022; Dehal et al., 2022; Sharma et al., 2020; Shen et al., 2017).

(VIII) There should be awareness programs and advertisement campaigns for making household people aware about the proper technique of disposal of BMW so that people cannot do random dumping of such hazardous waste. Household members should take an active part in such awareness activities (Fig. 3.2).

3.6 Conclusion

Amid the COVID-19 pandemic, the world has gone through environmental and health reformation that has impacted the world economy significantly. During the pandemic time, due to restricted movement and subsequent lockdowns leads to clean environment by restricting pollution. However, at the same time, the COVID-19 pandemic led to an increase in BMW generation in households and healthcare facilities that imposed hazardous effects on human health and the environment. Therefore, BMW emerged as a big threat to the world, especially to developing countries. Despite having current BMW management policies, it is necessary to revise the management policies on a common platform so that uniform policies can be executed around the world at the same time, as advised by the WHO. Post-COVID-19, there is also a need to reform our BMW system and replace the existing system with current technological advances.

Subsequently, the health workers and households should be trained for the disposal of such waste. Local administration should make stringent policies for reducing the volume of BMW generated at HCF by implementing the recycling policies and their alternatives.

Acknowledgments

The authors thank the Brain Pool Program funded by the Ministry of Science and ICT through the National Research Foundation of Korea (KRF Grant Number 2022H1D3A2A01096346) for supporting the research.

References

Acharya, A., Gokhale, V. A., & Joshi, D. (2014). Impact of biomedical waste on city environment: Case study of Pune, India. *Journal of Applied Chemistry, 6*(6), 21–27.

Agamuthu, P., & Barasarathi, J. (2021). Clinical waste management under COVID-19 scenario in Malaysia. *Waste Management & Research, 39*(1_suppl), 18–26.

Awoyera, P. O., & Adesina, A. (2020). Plastic wastes to construction products: Status, limitations and future perspective. *Case Studies in Construction Materials, 12*, e00330.

Bhar, A., Biswas, R. K., & Choudhury, A. K. (2022). The influence of COVID-19 pandemic on biomedical waste management: The impact beyond infection. *Proceedings of the Indian National Science Academy, 88*, 117–128.

Canopoli, L., Fidalgo, B., Coulon, F., & Wagland, S. T. (2018). Physico-chemical properties of excavated plastic from landfill mining and current recycling routes. *Waste Management, 76*, 55–67.

Capoor, M. R., & Parida, A. (2021). Biomedical waste and solid waste management in the time of COVID-19: A comprehensive review of the national and international scenario and guidelines. *Journal of Laboratory Physicians, 13*(02), 175–182.

CDC COVID-19 Response Team. (2020). Characteristics of health care personnel with COVID-19: United States. *MMWR Morbidity and Mortality Weekly Report, 69*, 477–481.

Chand, S., Shastry, C. S., Hiremath, S., Joel, J. J., Krishnabhat, C. H., & Mateti, U. V. (2021). Updates on biomedical waste management during COVID-19: The Indian scenario. *Clinical Epidemiology and Global Health, 11*, 100715.

Central Pollution Control Board (CPCB). (2016). Bio-Medical Waste Management Rules, 2016. Ministry of Environment, Forest and Climate Change, Government of India, New Delhi. Available at: Central Pollution Control Board – Bio-Medical Waste Management Rules, 2016 (Accessed: 3 July, 2026)

Chen, N., Zhou, M., Dong, X., Qu, J., Gong, F., Han, Y., Qiu, Y., Wang, J., Liu, Y., Wei, Y., & Yu, T. (2020). Epidemiological and clinical characteristics of 99 cases of 2019 novel coronavirus pneumonia in Wuhan, China: A descriptive study. *The Lancet, 395*(10223), 507–513.

Çobanoğlu, H., Belivermiş, M., Sıkdokur, E., Kılıç, Ö, & Çayır, A. (2021). Genotoxic and cytotoxic effects of polyethylene microplastics on human peripheral blood lymphocytes. *Chemosphere, 272*, 129805.

CPCB (2020) Guidelines for handling, treatment and disposal of waste generated during treatment/diagnosis/quarantine of COVID-19 patients, https://tspcb.cgg.gov.in/COVID19/COVID-19%20CPCB%20Guidelines%20 (Revision-4)%20for%20hanlding%20of%20COVID%20BM%20Waste%2021-07-2020.pdf

Cullen, W., Gulati, G., & Kelly, B. D. (2020). Mental health in the COVID-19 pandemic. *QJM: An International Journal of Medicine, 113*(5), 311–312.

Das, A. K., Islam, M. N., Billah, M. M., & Sarker, A. (2021). COVID-19 pandemic and health-care solid waste management strategy—A mini-review. *Science of the Total Environment, 778*, 146220.

Datta, P., Mohi, G. K., & Chander, J. (2018). Biomedical waste management in India: Critical appraisal. *Journal of Laboratory Physicians, 10*(1), 6–14.

Deepak, A., Kumar, D., & Sharma, V. (2021). Developing an effectiveness index for biomedical waste management in Indian states using a composite indicators approach. *Environmental Science and Pollution Research International, 28*(45), 64014–64029.

Dehal, A., Vaidya, A. N., & Kumar, A. R. (2022). Biomedical waste generation and management during COVID-19 pandemic in India: Challenges and possible management strategies. *Environmental Science and Pollution Research International, 29*(10), 14830–14845.

Dey, S., Anand, U., Kumar, V., Kumar, S., Ghorai, M., Ghosh, A., Kant, N., Suresh, S., Bhattacharya, S., Bontempi, E., & Bhat, S. A. (2023). Microbial strategies for degradation of microplastics generated from COVID-19 healthcare waste. *Environmental Research, 216*, 114438.

Dhaka, V., Singh, S., Anil, A. G., Sunil Kumar Naik, T. S., Garg, S., Samuel, J., Kumar, M., Ramamurthy, P. C., & Singh, J. (2022). Occurrence, toxicity and remediation of polyethylene terephthalate plastics: A review. *Environmental Chemistry Letters*, 1–24.

Fehr, A. R., & Perlman, S. C. (2015). An overview of their replication and pathogenesis. *Methods in Molecular Biology, 1282*, 1–23.

Gostin, L. O., Friedman, E. A., & Wetter, S. A. (2020). Responding to COVID-19: How to navigate a public health emergency legally and ethically. *Hastings Center Report, 50*(2), 8–12.

Gowda, N. R., Siddharth, V., Inquillabi, K., & Sharma, D. K. (2022). War on waste: Challenges and experiences in COVID-19 waste management. In *Disaster medicine and public health preparedness: 16* (pp. 2358–2362). Cambridge University Press.

Hwang, J., Choi, D., Han, S., Choi, J., & Hong, J. (2019). An assessment of the toxicity of polypropylene microplastics in human-derived cells. *Science of the Total Environment, 684*, 657–669.

Ilyas, S., Srivastava, R. R., & Kim, H. (2020). Disinfection technology and strategies for COVID-19 hospital and bio-medical waste management. *Science of the Total Environment, 749*, 141652.

Kampf, G., Todt, D., Pfaender, S., & Steinmann, E. (2020). Persistence of coronaviruses on inanimate surfaces and their inactivation with biocidal agents. *Journal of Hospital Infection, 104*(3), 246–251.

Khan, B. A., Cheng, L., Khan, A. A., & Ahmed, H. (2019). Healthcare waste management in Asian developing countries: A mini review. *Waste Management & Research, 37*(9), 863–875.

Khoo, K. S., Ho, L. Y., Lim, H. R., Leong, H. Y., & Chew, K. W. (2021). Plastic waste associated with the COVID-19 pandemic: Crisis or opportunity? *Journal of Hazardous Materials, 417*, 126108.

Khosla, R., Jha, A., Dua, S., Varmani, S. G., Rajput, N., & Pani, B. (2022). Upsurge in biomedical waste due to COVID-19 in India: A statistical correlation, challenges, and recommendations. *Frontiers in Environmental Science, 10*, 1022098. doi:10.3389/fenvs.2022.1022098.

Liu, J., Vethaak, A. D., An, L., Liu, Q., Yang, Y., & Ding, J. (2021). An environmental dilemma for China during the COVID-19 pandemic: The explosion of disposable plastic wastes. *Bulletin of Environmental Contamination and Toxicology, 106*(2), 237–240.

Lo, D., & De Angelis, M. (2020). COVID-19: Protecting health-care workers. *Lancet, 395*, 922.

Luhar, I., Luhar, S., & Abdullah, M. M. (2022). Challenges and impacts of COVID-19 pandemic on global waste management systems: A Review. *Journal of Composites Science, 6*(9), 271.

Manzoor, J., & Sharma, M. (2019). Impact of biomedical waste on environment and human health. *Environmental Claims Journal, 31*(4), 311–334.

Marinković, N., Vitale, K., Holcer, N. J., Džakula, A., & Pavić, T. (2008). Management of hazardous medical waste in Croatia. *Waste Management, 28*(6), 1049–1056.

Miyah, Y., Benjelloun, M., Lairini, S., & Lahrichi, A. (2022). COVID-19 impact on public health, environment, human, psychology, global socioeconomy, and education. *The Scientific World Journal, 2022*(1), 5578284.

Mohanan, P. V., Sangeetha, V., Sabareeswaran, A., Muraleedharan, V., Jithin, K., Vandana, U., & Varsha, S. B. (2021). Safety of 0.5% hydrogen peroxide mist used in the disinfection gateway for COVID-19. *Environmental Science and Pollution Research International, 28*(47), 66602–66612.

Ngoc, S. M., Nguyen, M. A., Nguyen, T. L., Thi, H. V., Dao, T. L., Bui, T. M., & Chu, D. T. (2022). COVID-19 and environmental health: A systematic analysis for the global burden of biomedical waste by this epidemic. *Case Studies in Chemical and Environmental Engineering, 6*, 100245.

Pillay, A. L., & Barnes, B. R. (2020). Psychology and COVID-19: Impacts, themes and way forward. *South African Journal of Psychology, 50*(2), 148–153.

Prata, J. C., Silva, A. L., Da Costa, J. P., Mouneyrac, C., Walker, T. R., Duarte, A. C., & Rocha-Santos, T. (2019). Solutions and integrated strategies for the control and mitigation of plastic and microplastic pollution. *International Journal of Environmental Research and Public Health, 16*(13), 2411.

Ramteke, S., & Sahu, B. L. (2020). Novel coronavirus disease 2019 (COVID-19) pandemic: Considerations for the biomedical waste sector in India. *Case Studies in Chemical and Environmental Engineering, 2*, 100029.

Ray, S. S., Lee, H. K., Huyen, D. T., Chen, S. S., & Kwon, Y. N. (2022). Microplastics waste in environment: A perspective on recycling issues from PPE kits and face masks during the COVID-19 pandemic. *Environmental Technology & Innovation, 26*, 102290.

Rowan, N. J., & Laffey, J. G. (2020). Challenges and solutions for addressing critical shortage of supply chain for personal and protective equipment (PPE) arising from Coronavirus disease (COVID-19) pandemic - Case study from the Republic of Ireland. *Science of the Total Environment, 725*, 138532.

Shammi, M., Behal, A., & Tareq, S. M. (2021). The escalating biomedical waste management to control the environmental transmission of COVID-19 pandemic: A perspective from two South Asian countries. *Environmental Science & Technology, 55*(7), 4087–4093.

Shammi, M., Bodrud-Doza, M., Towfiqul Islam, A. R. M., & Rahman, M. M. (2020). COVID-19 pandemic, socioeconomic crisis and human stress in resource-limited settings: A case from Bangladesh. *Heliyon, 6*(5), e04063.

Sharma, H. B., Vanapalli, K. R., Cheela, V. S., Ranjan, V. P., Jaglan, A. K., Dubey, B., Goel, S., & Bhattacharya, J. (2020). Challenges, opportunities, and innovations for effective solid waste management during and post COVID-19 pandemic. *Resources, Conservation & Recycling, 162*, 105052.

Shen, Y., Yu, S., Ge, S., Chen, X., Ge, X., & Chen, M. (2017). Hydrothermal carbonization of medical wastes and lignocellulosic biomass for solid fuel production from lab-scale to pilot-scale. *Energy, 118*, 312–323. doi:10.1016/j.energy.2016.12.047.

Siwal, S. S., Chaudhary, G., Saini, A. K., Kaur, H., Saini, V., Mokhta, S. K., & Thakur, V. K. (2021). Key ingredients and recycling strategy of personal protective equipment (PPE): Towards sustainable solution for the COVID-19 like pandemics. *Journal of Environmental Chemical Engineering, 9*, 106284.

The New York Times. Coronavirus updates: The illness now has a name, COVID-19. The New York Times.

Tsai, W. T. (2021). Analysis of medical waste management and impact analysis of COVID-19 on its generation in Taiwan. *Waste Management & Research, 39*(1_suppl), 27–33.

WHO, (2021). Health-care waste. https://www.who.int/news-room/fact-sheets/detail/health-care-waste.

WHO, (n.d.a), https://apps.who.int/iris/bitstream/handle/10665/259491/WHO-FWC-WSH-17.05-eng.pdf.

WHO, (n.d.b), https://www.unep.org/news-and-stories/story/healthcare-waste-what-do-it.

Vaughan, A. *11). The WHO still isn't describing COVID-19 as a pandemic. New Scientist.* https://www.newscientist.com/article/2235095-the-who-still-isnt-describing-covid-19-as-a-pandemic/.

World Health Organization (2020). Overview of public health and social measures in the context of COVID-19: Interim guidance. Geneva: World Health Organization. https://www.who.int/publications/i/item/WHO-2019-nCoV-PHSM_Overview-2020.1 (Accessed: 3 July, 2026).

Worldometer. Current world population. https://www.worldometers.info/world-population/. (Accessed 3 July, 2026).

Zhou, P., Yang, X. L., Wang, X. G., Hu, B., Zhang, L., Zhang, W., Si, H. R., Zhu, Y., Li, B., Huang, C. L., Chen, H. D., Chen, J., Luo, Y., Guo, H., Jiang, R. D., Liu, M. Q., Chen, Y., Shen, X. R., Wang, X., … Shi, Z. L. (2020). A pneumonia outbreak associated with a new coronavirus of probable bat origin. *Nature, 579*(7798), 270–273.

CHAPTER

Waste segregation techniques 4

Raj Morya [a], Vivek Morya [b], Abhay Raj [c] and Sunil Kumar [d]

[a] *Postdoctoral Researcher, Laboratory of Environmental Biotechnology (LBE), National Research Institute for Agriculture, Food and Environment (INRAE), Narbonne, France,* [b] *Department of Orthopedic Surgery, Dongtan Sacred Heart Hospital Dongtan, Hallym University, Hwaseong-si, Gyeonggi-do, Republic of Korea,* [c] *Environmental Microbiology Division, CSIR-Indian Institute of Toxicology Research, Vishvigyan Bhavan 31, Mahatma Gandhi Marg, Lucknow, Uttar Pradesh, India,* [d] *Centre for Industrial Ecology and Circular Economy, CSIR-National Environmental Engineering Research Institute, Nagpur, Maharashtra, India*

4.1 Introduction

Waste segregation forms the backbone of modern waste management systems, ensuring that waste streams are systematically classified, handled, and disposed of in ways that minimize environmental harm. By separating waste at its origin, communities and institutions can significantly reduce health hazards, conserve landfill capacity, and facilitate recycling and resource recovery. Beyond environmental benefits, proper segregation enhances regulatory compliance, reduces costs, and fosters public awareness of sustainable practices. This chapter explores the importance of waste segregation at source, highlights best practices for biomedical waste management, and discusses tools, equipment, and technologies essential for effective implementation.

4.2 Importance of waste segregation at source

Source-level segregation is the single most critical step in achieving effective waste management. By classifying waste at the point of generation into categories such as hazardous, recyclable, biodegradable, and nonrecyclable, entire waste management chains—from collection to treatment—become more efficient and sustainable (Gamaralalage et al., 2022; Kumawat et al., 2024). The practice not only reduces risks to human and environmental health but also optimizes economic and regulatory outcomes.

4.2.1 Reduction in landfill dependency

Segregating waste at the source diverts a significant fraction away from landfills. Recyclables like paper, plastics, and metals can reenter production cycles, while biodegradable waste can be composted or converted into biogas. This reduces methane

Biomedical Waste Management in the COVID-19 Era: A Global Scenario.
DOI: https://doi.org/10.1016/B978-0-443-21841-5.00010-2

emissions, prolongs the lifespan of landfill sites, and curtails environmental degradation.

4.2.2 Improved recycling efficiency

The success of recycling systems depends heavily on the quality of input materials. Clean, segregated waste enhances recycling efficiency, reduces operational costs at treatment plants, and increases recovery rates of valuable materials. In contrast, contaminated mixed waste often ends up in landfills despite its recyclable potential.

4.2.3 Minimization of health risks

Unsegregated waste often contains hazardous fractions such as biomedical residues, chemicals, and electronic discards. These pose serious risks to sanitation workers, healthcare professionals, and the general public. Proper segregation acts as a preventive barrier against infections, injuries, and toxic exposure.

4.2.4 Cost-effective waste management

Sorting waste at the point of generation reduces downstream processing costs. Collection and transport systems are streamlined when recyclable and organic fractions are separated, leaving only residual nonrecyclables for final disposal. Municipalities adopting segregation have reported significant cost savings in waste management infrastructure.

4.2.5 Environmental protection and pollution reduction

Segregation prevents hazardous materials from contaminating soil, water, and air. For example, keeping e-waste separate curtails heavy metal leaching into groundwater, while isolating organics from landfills reduces methane release—a greenhouse gas with 28 times the warming potential of CO_2.

4.2.6 Compliance with regulations and global goals

Waste segregation is mandated by multiple national and international frameworks, including India's BMWM bio-medical waste management (BMWM) Rules 2016 and the UN [Sustainable Development Goals (SDG) 12: Responsible Consumption and Production; SDG 13: Climate Action]. Compliance not only avoids penalties but also strengthens institutional accountability and environmental stewardship.

4.2.7 Promotion of public responsibility and awareness

Segregation fosters civic engagement and environmental responsibility. Households and organizations become more mindful of consumption and disposal habits when visible segregation practices are enforced. Educational campaigns and community participation further embed a culture of sustainability (Table 4.1).

Table 4.1 Key benefits of waste segregation at the source.

Benefit	Example/impact
Reduced landfill dependency	Biodegradables diverted to compost/biogas; extended landfill life
Improved recycling efficiency	Higher recovery of uncontaminated plastics, paper, and metals
Health risk reduction	Fewer needlestick injuries; reduced exposure to toxins
Cost savings	Lower processing and transport costs
Environmental protection	Mitigated groundwater pollution, reduced methane emissions
Regulatory compliance	Alignment with BMWM rules, EPA standards, and UN SDGs
Public awareness	Greater environmental consciousness and behavioral change

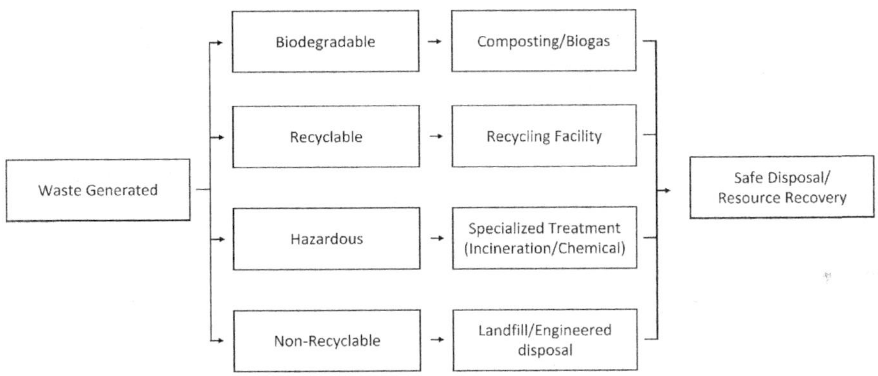

FIGURE 4.1

Flow of waste segregation at the source.

4.3 Best practices for segregating different types of biomedical waste

Biomedical waste contains a wide range of hazardous materials that, if improperly managed, can endanger both human health and the environment. Effective segregation is therefore a critical step in ensuring safe disposal, preventing contamination, and meeting both national and international regulatory requirements (China et al., 2024; Cook et al., 2023; WHO, 2022). Proper segregation reduces the risk of infections, accidental injuries, and environmental pollution, while also improving the efficiency of downstream treatment and disposal processes (Fig. 4.1).

Table 4.2 Biomedical waste categories and color-coded segregation.

Waste category	Color code	Examples	Disposal method
Infectious waste	Yellow	Body fluids, human tissues, soiled bandages, and gloves	Autoclaving or incineration
Sharps waste	Red	Syringes, scalpels, broken glass, and needles	Puncture-proof containers → sterilization or incineration
Pharmaceutical waste	Blue	Unused medicines, vaccines, and expired drugs	Incineration or return to authorized supplier
Cytotoxic waste	Purple	Chemotherapy waste and cancer treatment drugs	High-temperature incineration (>1200°C) in specialist facilities
General nonhazardous	Black	Food waste, paper, packaging materials, and plastics	Municipal waste stream (recycling/landfill, depending on category)

Sources: China et al. (2024); WHO (2022); Bansod et al. (2023); Cho et al., 2024; Cook et al. (2023).

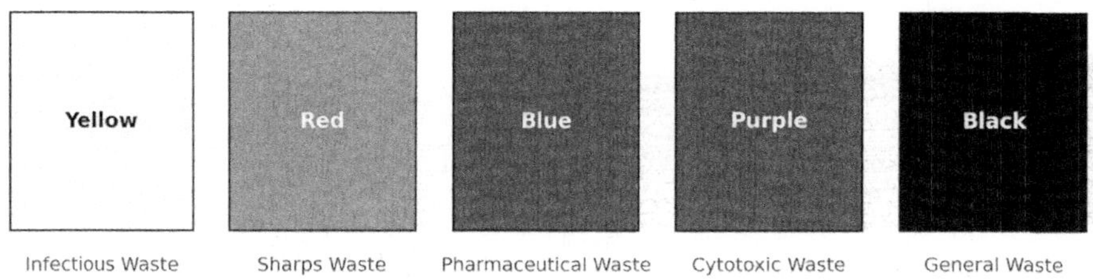

FIGURE 4.2

Example of color-coded bins.

4.3.1 Color-coded waste segregation system

A widely adopted method for biomedical waste management is the color-coded segregation system, which ensures clear categorization and safe disposal of specific waste streams (Bansod et al., 2023; China et al., 2024). This system prevents hazardous and nonhazardous waste from mixing, thereby minimizing occupational risks and streamlining treatment processes (Table 4.2).

4.3.2 Handling and storage guidelines

Effective biomedical waste management extends beyond segregation. Proper handling and storage practices are essential to maintain safety and regulatory compliance based on (Cook et al., 2023; Sharior et al., 2023) (Fig. 4.2).

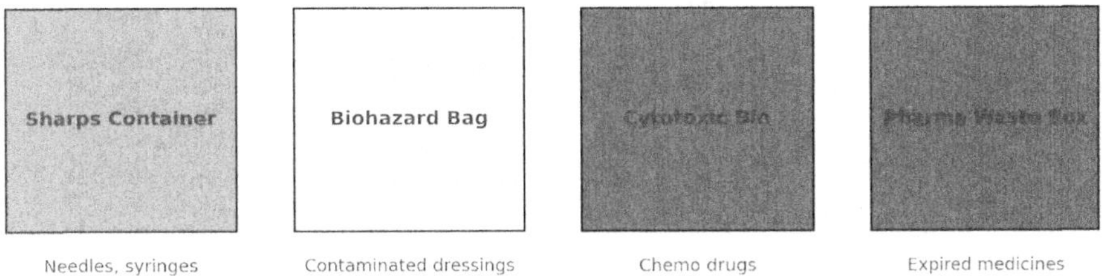

FIGURE 4.3

Specialized biomedical waste containers.

- *Training and awareness*: All healthcare staff, janitors, and waste handlers should undergo regular training sessions on waste segregation, handling, and safety precautions (Gebreeyessus, 2022). Refresher courses and workshops foster accountability and compliance.
- *Labeling and containment*: Waste bags and containers must be clearly labeled with text and internationally recognized symbols. Sharps should be stored in puncture-proof, tamper-resistant containers, while infectious and cytotoxic waste must be sealed in leak-proof bags (Sharior et al., 2023).
- *Storage conditions*: Waste should be stored in ventilated areas, away from patient care zones. Infectious waste should not remain on-site for more than 48 h before transport to treatment facilities (Cook et al., 2023).
- *Safety and hygiene measures*: Staff handling waste must wear personal protective equipment (PPE) such as gloves, masks, aprons, and face shields (WHO, 2022). Hand hygiene is essential, and spills must be cleaned immediately with appropriate disinfectants (Slutzman et al., 2023) (Fig. 4.3).

4.4 Personal protective equipment

PPE forms the first line of defense for healthcare workers and waste handlers against hazards associated with biomedical waste. Proper use of PPE reduces exposure to infectious agents, hazardous chemicals, and sharp objects, thereby safeguarding occupational health and ensuring compliance with national and international safety guidelines (Persaud et al., 2023; WHO, 2022).

The choice of PPE must adhere to standardized performance indicators, material specifications, and regulatory compliance frameworks such as EN 455 (Europe), ASTM D6319 (USA), NIOSH (42 CFR 84, USA), FDA (21 CFR 878.4040, USA), and ISO 16603/16604 (International) (Freitas et al., 2025; Moore, Coffey, Duling, & D'Alessandro, 2024). Each component has distinct protective functions, as summarized in Table 4.3.

PPE, personal protective equipment.

Table 4.3 Key PPE components for biomedical waste management.

PPE component	Key features	Regulatory standards	Protective role
Gloves (latex/nitrile/vinyl)	Nitrile/latex offers a superior barrier; vinyl is prone to higher failure rates	EN 455 (EU), ASTM D6319 (USA)	Prevent contact with bloodborne pathogens, chemicals, and contaminated surfaces
Face masks and respirators	N95: ≥95% filtration of 0.3 μm particles; surgical masks with ≥95% BFE	NIOSH 42 CFR 84, FDA 21 CFR 878.4040	Protect against airborne pathogens, splashes, and droplets
Face shields	Wraparound, polycarbonate/PETG material, impact-resistant	ANSI Z87.1 (USA), EN 166 (EU)	Splash and droplet protection for eyes and face
Protective clothing (aprons and coveralls)	Fluid-resistant, impermeable (e.g., Tyvek and PE-coated polypropylene)	ASTM F1670/F1671, ISO 16603/16604	Prevent penetration of blood, body fluids, and viruses
Goggles	Chemical- and impact-resistant polycarbonate with indirect venting	ANSI Z87.1, EN 166	Protect eyes from infectious splashes and chemical spills
Footwear	Impermeable boots, closed-toe, and slip-resistant; may include puncture-resistant soles	ASTM F2413 (USA), EN ISO 20345 (EU)	Prevent fluid infiltration, sharps injuries, and slips

FIGURE 4.4

Personal protective equipment (PPE) in biomedical waste management.

Key considerations

1. *Gloves:* Latex and nitrile are preferred over vinyl due to superior integrity; standards specify acceptable failure rates ≤4% under simulated clinical use (Freitas et al., 2025; Herkins et al., 2024).
2. *Masks/respirators*: N95 respirators must be NIOSH-certified, while surgical N95s also require FDA approval for fluid resistance (Lindsley et al., 2021; Moore, Coffey, Duling, & D'Alessandro, 2024).
3. *Protective clothing*: Coveralls/aprons should resist viral penetration as per ASTM F1671, with sealed seams and full-body coverage (Morris & Murray, 2021).
4. *Eye protection*: Goggles must offer lateral coverage and compatibility with masks/shields (Fogarasi et al., 2023).
5. *Footwear*: Impermeable, slip-resistant boots minimize risks of sharps injury and contamination (Morris & Murray, 2021) (Fig. 4.4).

Proper donning, doffing, and disposal protocols are critical to ensuring PPE effectiveness. Training sessions, fit testing of respirators, and regular quality checks are recommended for all healthcare facilities (Gebreeyessus, 2022; WHO, 2022).

4.5 Segregation and monitoring technologies

The growing complexity of biomedical waste streams necessitates adopting technology-driven solutions to ensure safe, efficient, and transparent segregation, tracking, and treatment. These innovations reduce reliance on manual handling, minimize occupational risks, and enhance regulatory compliance (Araujo-Andrade et al., 2021; Liu et al., 2024).

4.5.1 Automated waste sorting systems

Traditional biomedical waste segregation relies heavily on manual labor, which is time-consuming, error-prone, and exposes workers to infectious hazards. Automated waste sorting systems overcome these limitations by combining robotics, computer vision, hyperspectral imaging, and machine learning algorithms to classify waste streams in real time (Araujo-Andrade et al., 2021). High-resolution cameras and near-infrared (NIR) spectroscopy detect shape, color, and material composition, while artificial intelligence (AI) algorithms compare this data against trained models to distinguish plastics, metals, glass, organic matter, and hazardous components. Once identified, robotic arms or pneumatic ejectors remove target items from conveyor belts, significantly reducing contamination in recyclable fractions. The efficiency gains are notable: Automated systems not only improve segregation accuracy but also increase throughput in material recovery facilities (MRFs), lowering operational costs and minimizing occupational risks for waste handlers (Liu et al., 2024). For biomedical waste, where improper mixing of sharps, pharmaceuticals, and infectious materials can have severe consequences, these systems provide a scalable, safe, and resource-efficient alternative.

4.5.2 RFID and barcode tracking devices

Traceability of biomedical waste is essential for both regulatory compliance and environmental safety. Barcode labeling remains one of the most widely adopted methods due to its affordability and ease of integration into existing workflows. Each waste bag or container is labeled with a unique barcode, scanned at collection and transfer points, thereby recording its origin, category, and movement history. However, RFID technology offers a significant upgrade. Passive or active RFID chips embedded in containers can be read wirelessly at a distance, enabling real-time, contactless monitoring (Wawale et al., 2022). Unlike barcodes, RFID tags do not require line-of-sight scanning, which speeds up operations and reduces handling. Data such as location, weight, and time stamps are transmitted to centralized monitoring systems, strengthening accountability and minimizing risks of illegal dumping or unauthorized reuse of hazardous materials (Frith, 2024). For hospitals, RFID systems can also integrate with GPS-tracked transport vehicles, providing cradle-to-grave traceability of waste.

4.5.3 Digital waste auditing

Waste auditing is a vital tool for understanding generation patterns, segregation effectiveness, and disposal efficiency. Digital waste auditing systems enhance this process by using IoT sensors, cloud platforms, and advanced data analytics to provide continuous, real-time insights. Sensors placed in bins or collection areas record fill levels, material types, and contamination rates, while cloud-based dashboards aggregate this information into performance metrics and key performance indicators (KPIs) (Slutzman et al., 2023). These systems enable facility managers to identify "hotspots" of poor segregation, optimize collection schedules, and design targeted staff training programs. Moreover, automated compliance reports can be generated for submission to environmental and health authorities, ensuring regulatory adherence (Thanigaivelu et al., 2024). In biomedical settings, digital auditing also supports quality assurance, reducing risks of infectious waste being mishandled and strengthening overall hospital sustainability strategies.

4.5.4 Blockchain-based traceability

Blockchain technology introduces a decentralized, tamper-proof digital ledger that records every step of biomedical waste management, from generation to final disposal. Each transaction—such as waste collection, transport, treatment, or incineration—is time-stamped and cryptographically secured. Once entered, records cannot be altered retroactively, thereby preventing falsification of data or illegal diversion of hazardous waste (Wang et al., 2022). When integrated with RFID tags, QR codes, and IoT devices, blockchain enables full lifecycle traceability. For example, RFID-tagged waste bags can automatically upload weight, origin, and treatment data to the blockchain, ensuring transparency for regulators, hospitals, and waste operators. Furthermore, smart contracts can be programmed to enforce compliance, such as ensuring waste is treated within 48 h of generation or transported only to certified facilities (Bułkowska et al., 2023). By guaranteeing data integrity and fostering trust among stakeholders, blockchain has the potential to become a cornerstone of future biomedical waste governance, reducing risks of illegal dumping while enhancing public and environmental safety (Table 4.4).

4.6 Environmental responsibility and regulatory compliance

The safe management of biomedical waste is not only a matter of operational efficiency but also a legal and ethical obligation. Robust environmental responsibility ensures that hazardous waste does not compromise public health, ecosystems, or climate stability. To achieve this, strict national and international regulatory frameworks provide detailed guidelines for waste segregation, transportation, treatment, and disposal. Compliance with these frameworks mitigates risks, enhances sustainability, and builds institutional accountability. Biomedical waste is particularly dangerous due to its potential to spread infections, release toxic emissions, and contaminate natural resources if mismanaged. Thus, healthcare facilities, laboratories, and industries are

Table 4.4 Emerging technologies in biomedical waste segregation and monitoring.

Technology	Distinctive feature	Added value beyond manual methods	Example in practice	References
Automated sorting systems	AI-driven recognition of material properties (color, shape, spectral signature)	Enhances the purity of recyclable fractions; enables large-scale processing	Robotic removal of sharps/plastics from mixed hospital waste	Araujo-Andrade et al. (2021); Liu et al. (2024)
RFID and barcode tracking	Unique digital identity for each waste unit	Ensures cradle-to-grave accountability; reduces "leakage" into municipal waste	RFID-tagged bags transmitting live transport data	Wawale et al. (2022); Frith (2024)
Digital waste auditing	Sensor-based quantification and dashboard visualization	Provides real-time metrics (e.g., fill levels, contamination rate); supports decision-making	IoT-enabled bins generating compliance reports	Slutzman et al. (2023); Thanigaivelu et al. (2024)
Blockchain-based traceability	Immutable decentralized ledger with smart contracts	Guarantees data integrity; automates compliance enforcement	Blockchain record of cytotoxic waste incineration within 48 h	Wang et al. (2022); Bułkowska et al. (2023)

Table 4.5 Key regulatory frameworks governing biomedical waste management.

Regulatory framework	Jurisdiction	Core requirements	Environmental/ health safeguards	References
Biomedical Waste Management Rules (2016)	India	Segregation at source, barcoding, max 48 h storage, GPS-tracked transport, and Common Bio-medical Waste Treatment Facility (CBWTF) treatment	Prevents open dumping/burning, enhances traceability, and occupational safety	Singhal et al. (2017); Pande et al. (2024)
Environmental Protection Agency (EPA)—RCRA and CAA	USA	Hazardous waste handling under Resource Conservation and Recovery Act (RCRA) ; strict incinerator emissions under CAA	Controls dioxins, furans, and particulate emissions; promotes safe disposal	Rau et al. (2000); Rettze and Stagg (1998)
Occupational Safety and Health Administration (OSHA)	USA	PPE standards, sharps disposal, staff training, and hygiene requirements	Protects workers from needlestick injuries, exposure to pathogens, and chemicals	Persaud et al. (2023); Patra and Neithiya (2023)
World Health Organization (WHO) Guidelines	Global	Safe management of healthcare waste, PPE use, and environmentally sound disposal	Reduces infection risks, promotes sustainable technologies	WHO (2022)
European Union Waste Framework Directive	EU	Waste hierarchy (reduce, reuse, recycle, and recover), hazardous waste classification, and traceability	Promotes circular economy, harmonized EU-wide compliance	Bułkowska et al. (2023)

mandated to adhere to binding regulations such as India's Biomedical Waste Management Rules (2016), the US EPA standards, and occupational safety protocols under the OSHA. Each of these frameworks emphasizes segregation at source, traceability, PPE use, and sustainable treatment technologies. Importantly, these regulations align with international sustainability frameworks, particularly the United Nations SDGs, including SDG 3 (Good Health and Well-being), SDG 6 (Clean Water and Sanitation), SDG 12 (Responsible Consumption and Production), and SDG 13 (Climate Action). By enforcing these policies, biomedical waste management contributes directly to environmental protection and public health security (Table 4.5).

4.7 Challenges in biomedical waste segregation and disposal

Despite significant technological advances and increasingly stringent regulatory frameworks, the effective segregation and disposal of biomedical waste continue to face critical challenges. These barriers undermine safe waste handling, compliance with environmental norms, and sustainability goals. The following subsections highlight the major issues.

4.7.1 Lack of awareness and training

One of the most persistent challenges is the inadequate training of healthcare personnel regarding biomedical waste management. Studies show that improper knowledge of segregation protocols often leads to cross-contamination and noncompliance (Gebreeyessus, 2022; Sharior et al., 2023). Inadequately trained staff may dispose of sharps with general waste, mix hazardous with nonhazardous materials, or fail to follow color-coded bin systems, exposing both workers and patients to infection risks.

4.7.2 Inadequate infrastructure in developing regions

Many healthcare facilities, particularly in developing nations, lack access to reliable waste treatment technologies such as autoclaves, incinerators, or secured landfills (Chisholm et al., 2021). In rural and resource-constrained hospitals, open dumping and uncontrolled burning are still practiced, contributing to air pollution, water contamination, and community health hazards (Hantoko et al., 2021).

4.7.3 Illegal disposal and black-market risks

Illegal dumping of biomedical waste into municipal systems remains a serious issue. Hazardous components—including syringes, vials, and scalpels—are sometimes recovered and reused through informal markets, increasing the risk of disease outbreaks (Pindi et al., 2023; Sinha et al., 2020). Such practices not only endanger public health but also erode public trust in healthcare systems.

4.7.4 High cost of advanced technologies

While novel solutions such as plasma pyrolysis, IoT-based tracking, and AI-driven sorting hold promise, their high capital and operational costs restrict adoption in small- and medium-sized healthcare facilities (Jiang et al., 2022; Liu et al., 2024). The financial burden often compels institutions to rely on outdated methods that compromise both safety and sustainability.

4.8 Future directions in biomedical waste segregation

Innovative technologies and sustainable practices that seek to lessen the negative effects that conventional waste disposal methods have on the environment and public health are increasingly influencing the future directions of biomedical waste segregation. The creation of biodegradable medical supplies, such as gloves, syringes, and PPE, is one of the most promising developments. This field of study is quickly growing and aims to address the growing issue of plastic waste in healthcare settings. Through microbial activity and environmental exposure, these products are designed to decompose naturally over time, greatly minimizing their environmental impact, relieving pressure on overburdened waste management systems, and lowering the risk of long-term pollution. Biodegradable substitutes for traditional plastic-based medical devices, which can linger in landfills or the ocean for decades, provide environmentally friendly disposal options that are consistent with the concepts of environmental sustainability and the circular economy. By using these materials, healthcare organizations can lower their carbon footprint, meet sustainability targets, and actively support international environmental conservation initiatives. The use of AI-driven waste prediction models, which are transforming the way hospitals and clinics handle biomedical waste, complements this move toward sustainable materials. To predict the type and amount of waste that is likely to be generated, these models use machine learning algorithms to analyze large datasets, including patient flow, treatment types, and historical waste volumes. These technologies facilitate data-informed decision-making, which in turn leads to more precise waste segregation, improved resource allocation, and the adoption of proactive waste generation reduction methods, all of which contribute to cleaner, safer, and more effective healthcare environments. Furthermore, the creation and application of cutting-edge waste-to-energy technologies are showing promise as effective means of managing biomedical waste in an environmentally friendly way. Hazardous biological waste is transformed into renewable energy forms, including biogas, bio-oil, and electricity, using processes like pyrolysis, gasification, and anaerobic digestion. In addition to decreasing environmental risks and diverting waste from landfills, these technologies also aid in energy recovery, promoting a circular economy and lowering reliance on fossil fuels. These developments in energy recovery, biodegradable materials, and AI-driven analytics collectively offer a progressive, integrated method of biomedical waste segregation that has the potential to completely reshape sustainability norms in healthcare systems around the globe.

4.9 Conclusion

Segregation of biomedical waste is essential for maintaining a healthy environment and protecting public health. Waste management procedures have been completely transformed by the combination of blockchain, AI, IoT, and plasma pyrolysis. Nonetheless, issues including low knowledge, poor infrastructure, and exorbitant expenses must be resolved. Healthcare organizations may guarantee safer and more sustainable biological waste management by implementing best practices, making investments in innovative technologies, and enforcing stringent laws.

References

Araujo-Andrade, C., Bugnicourt, E., Philippet, L., Rodriguez-Turienzo, L., Nettleton, D., Hoffmann, L., & Schlummer, M. (2021). Review on the photonic techniques suitable for automatic monitoring of the composition of multi-materials wastes in view of their posterior recycling. *Waste Management & Research, 39*(5), 631–651.

Bansod, H. S., Deshmukh, P., & Bansod, H. (2023). Biomedical waste management and its importance: A systematic review. *Cureus, 15*(2), e34589'.

Bułkowska, K., Zielińska, M., & Bułkowski, M. (2023). Implementation of blockchain technology in waste management. *Energies, 16*(23), 7742.

China, M. A., Amadi, G. A., & Wachukwu, C. H. (2024). Utilization of colour-coded bins for management of solid waste among households in Obio Akpor Local Government Area, Rivers State. *Journal of Home Economics Research, 31*(2), 155–165.

Chisholm, J. M., Zamani, R., Negm, A. M., Said, N., Abdel Daiem, M. M., Dibaj, M., & Akrami, M. (2021). Sustainable waste management of medical waste in African developing countries: A narrative review. *Waste Management & Research, 39*(9), 1149–1163.

Cho, Y., Withana, P. A., Rhee, J. H., Lim, S. T., Lim, J. Y., Park, S. W., & Ok, Y. S. (2024). Achieving the sustainable waste management of medical plastic packaging using a life cycle assessment approach. *Heliyon, 10*(19), e38185.

Cook, E., Woolridge, A., Stapp, P., Edmondson, S., & Velis, C. A. (2023). Medical and healthcare waste generation, storage, treatment, and disposal: A systematic scoping review of risks to occupational and public health. *Critical Reviews in Environmental Science and Technology, 53*(15), 1452–1477.

Dickella Gamaralalage, P. J., Ghosh, S. K., & Onogawa, K. (2022). Source separation in municipal solid waste management: Practical means to its success in Asian cities. *Waste Management & Research, 40*(3), 360–370.

Fogarasi, M., Snodderly, K., Herman, A., Guha, S., & Porter, D. (2023). Benchtop assessment of sealing efficacy and breathability of additively manufactured (AM) face masks. *Additive Manufacturing, 67*, 103468.

Freitas, J., Lomba, A., Sousa, S., Gonçalves, V., Brois, P., Nunes, E., Veloso, I., Peres, D., & Alves, P. (2025). Consensus-based guidelines for best practices in the selection and use of examination gloves in healthcare settings. *Nursing Reports, 15*(1), 9.

Frith, J. (2024). *A billion little pieces: RFID and infrastructures of identification*. Mit Press.

Gebreeyessus, G. D. (2022). Knowledge, attitude, and practices on occupational health and safety principles among cleaners: The case of Tikur Anbassa specialized referral hospital, Addis Ababa, Ethiopia. *Open Health, 3*(1), 22–33.

Hantoko, D., Li, X., Pariatamby, A., Yoshikawa, K., Horttanainen, M., & Yan, M. (2021). Challenges and practices on waste management and disposal during COVID-19 pandemic. *Journal of Environmental Management, 286*, 112140.

Herkins, A., Dey, S., Conroy, D., & Cornish, K. (2024). Nitrile glove composition and performance—Substandard properties and inaccurate packaging information. *PLoS ONE, 19*(10), e0312891.

Huang, B., Gan, M., Ji, Z., Fan, X., Zhang, D., Chen, X., Sun, Z., Huang, X., & Fan, Y. (2022). Recent progress on the thermal treatment and resource utilization technologies of municipal waste incineration fly ash: A review. *Process Safety and Environmental Protection, 159*, 547–565.

Imam, A. B. (2021). *Blockchain-enabled approach for healthcare waste management.* Khalifa University of Science.

Jiang, X., Zhao, Y., & Yan, J. (2022). Disposal technology and new progress for dioxins and heavy metals in fly ash from municipal solid waste incineration: A critical review. *Environmental Pollution, 311*, 119878.

Kharat, D. S. (2016). Biomedical waste management rules, 2016: A review. *International Journal of Advanced Research and Development, 1*, 48–51.

Kumawat, A. K., Vaish, S., & Pathak, B. (2024). Understanding waste: Type, characterization, and environmental impacts. In *Solid waste management* (pp. 1–25). CRC Press.

Lindsley, W. G., Blachere, F. M., Law, B. F., Beezhold, D. H., & Noti, J. D. (2021). Efficacy of face masks, neck gaiters and face shields for reducing the expulsion of simulated cough-generated aerosols. *Aerosol Science and Technology, 55*(4), 449–457.

Liu, M., Grimes, S., Salonitis, K., & Litos, L. (2024). Modular model and simulation for process optimisation in advanced material recovery facilities (MRFs). *Procedia CIRP, 128*, 250–255.

Moore, S. M., Coffey, C., Duling, M., & D'Alessandro, M. (2024). *NIOSH healthcare personal protective technology targets for 2020 to 2030.Pittsburgh.* PA: U.S. Department of Health and Human Services, Centers for Disease Control and Prevention, National Institute for Occupational Safety and Health. DHHS https://doi.org/10.26616/NIOSHPUB2024107.

Morris, H., & Murray, R. (2021). Healthcare, hygiene, and personal protective equipment (PPE). In *Medical textiles* (pp. 261–310). CRC Press.

Ogunkan, D. V., & Ogunkan, S. K. (2025). Exploring big data applications in sustainable urban infrastructure: A review. *Urban Governance, 5*(1), 54–68.

Pande, H., Varughese, J. C., Singh, K., Sonecha, S., Prajapati, D., Joshi, K., Kumar, H., Bhutani, S., & Singh, M. (2024). Segregation of biomedical waste: Methodologies and importance. *Biomedical Waste Management: Bioremediation and Recycling, 1*, 65–84.

Patra, A. P., & Neithiya, T. (2023). Basics of healthcare facilities management. In *Medical jurisprudence & clinical forensic medicine* (pp. 201–207). CRC Press.

Persaud, E., Parker, G. B., & Mitchell, A. H. (2023). Blood and body fluid exposure among healthcare workers and personal protective equipment usage in the United States. *Workplace Health & Safety, 71*(9), 429–435.

Pindi, G., Suravaram, S., Kandluri, S., & Donavalli, K. (2023). Biomedical waste segregation compliance scoring system: To analyze, strengthen, monitor, and step up waste management in healthcare facilities. *Pan African Medical Journal, 45*(1).

Rau, E. H., Alaimo, R. J., Ashbrook, P. C., Austin, S. M., Borenstein, N., Evans, M. R., French, H. M., Gilpin, R. W., Hughes, J., Hummel, S. J., Jr, Jacobsohn, A. P., Lee, C. Y., Merkle, S., Radzinski, T., Sloane, R., Wagner, K. D., & Weaner, L. E. (2000). Minimization and management of wastes from biomedical research. *Environmental Health Perspectives, 108*(suppl 6), 953–977.

Rettze, A. W., Jr, & Stagg, M. K. (1998). Air emissions standards and guidelines under the Clean Air Act for the incineration of hospital, medical, and infectious waste. *Environmental Law, 28*, 791.

Sharior, F., Alam, M. U., Zaqout, M., Cawood, S., Ferdous, S., Shoaib, D. M., Tidwell, J. B., Hasan, M., Hasan, M., Rahman, M., Farah, M., Rahman, M. A., Ahmed, A., & Ahmed, T. (2023). Occupational health and safety status of waste and sanitation workers: A qualitative exploration during the COVID-19 pandemic across Bangladesh. *PLOS Water, 2*(1), e0000041.

Singhal, L., Tuli, A. K., & Gautam, V. (2017). Biomedical waste management guidelines 2016: What's done and what needs to be done. *Indian Journal of Medical Microbiology, 35*(2), 194–198.

Sinha, S., Babu, N. A., Behura, S. S., & Rajesh, E. (2020). Management of biomedical waste in hospital and health care organizations: A review. *Indian Journal of Forensic Medicine & Toxicology, 14*(4), 1256–1260.

Slutzman, J. E., Bockius, H., Gordon, I. O., Greene, H. C., Hsu, S., Huang, Y., Lam, M. H., Roberts, T., & Thiel, C. L. (2023). Waste audits in healthcare: A systematic review and description of best practices. *Waste Management & Research, 41*(1), 3–17.

Thanigaivelu, P. S., Lakshmi, D., Ramasamy, J., Kumaresh, P. S., Sumathi, G., & Mathivanan, K. (2024). IoT and cloud solutions for waste-to-energy in hospitals for energy recovery from medical waste. In *2024 10th International Conference on Communication and Signal Processing (ICCSP)* (pp. 1112–1117). IEEE.

Wang, H., Zheng, L., Xue, Q., & Li, X. (2022). Research on medical waste supervision model and implementation method based on blockchain. *Security and Communication Networks, 2022*(1), 5630960.

Wawale, S. G., Shabaz, M., Mehbodniya, A., Soni, M., Deb, N., Elashiri, M. A., & Naved, M. (2022). Biomedical waste management using IoT-tracked and fuzzy-classified integrated technique. *Human-Centric Computing and Information Sciences, 12*(32), 32.

World Health Organization. (2022). *Global analysis of healthcare waste in the context of COVID-19: Status, impacts, and recommendations*. World Health Organization.

CHAPTER

5 Biomedical waste management in the post-COVID era: Global and South Asian perspectives

Vivek Kumar Morya

Department of Orthopedic Surgery, Hallym University Sacred Heart Hospital, Anyang-si, Gyeonggi-do, Republic of Korea

5.1 Introduction

Biomedical waste management (BMW) has emerged as a critical component of public health infrastructure in the post-COVID era. *Biomedical waste* refers to any waste generated during the diagnosis, treatment, or immunization of humans or animals, including research activities and production/testing of biologicals (World Health Organization 2022). This waste stream encompasses materials that may be infectious, toxic, or otherwise hazardous, such as used syringes, swabs, bandages, discarded medicines, laboratory samples, and personal protective equipment (PPE). Safe handling and disposal of BMW is essential because improper waste management can expose healthcare workers, waste handlers, patients, and the community to infection or injury. In fact, the World Health Organization (WHO) estimates that about 15% of healthcare waste is hazardous (infectious, chemical, or radioactive), while the remaining 85% is general nonhazardous waste (Manekar et al., 2022). However, in many low-income settings, lack of segregation leads to commingling of waste, effectively making a much larger fraction of waste hazardous (Faisal et al., 2021).

The COVID-19 pandemic has significantly magnified the challenges of BMW worldwide. The massive scale of the pandemic response, including testing, vaccination, and the use of single-use PPE, generated unprecedented quantities of medical waste. For example, during the height of the outbreak in Wuhan, China, hospitals produced **over 240 tons of medical waste per day**, compared to about 40 tons per day before the pandemic (Luhar et al., 2022). Similar surges were observed globally: Cities like Manila, Kuala Lumpur, Hanoi, and others generated an **additional 154 280 tons of healthcare waste per day** compared to prepandemic times [Central Pollution Control Board (India) 2020]. This sudden influx of infectious waste strained waste management systems, many of which were already under resourced. The situation has been especially acute in developing regions, including South Asia, where baseline waste management capacity was limited, and **30% of healthcare facilities (and**

Biomedical Waste Management in the COVID-19 Era: A Global Scenario.
DOI: https://doi.org/10.1016/B978-0-443-21841-5.00008-4

as many as 60% in least-developed countries) lacked adequate waste handling systems even before COVID-19 (World Health Organization 2020).

The COVID-19 pandemic not only tested healthcare systems through patient care; it quietly but profoundly reshaped how biomedical waste was generated, handled, and disposed of across the world. This chapter explores that transformation, tracing the journey of BMW from the height of the crisis to the evolving postpandemic landscape, with particular attention to South Asia alongside global experiences. It moves through the everyday realities of hospitals, the urgency of national policy responses, and the influence of international regulatory guidance. Along the way, it reflects on how long-standing waste management practices were stretched to their limits, forcing institutions to adapt quickly and, in many cases, rethink established routines. The chapter also highlights the innovations that emerged from this pressure, new methods of treatment, tracking, and disposal that are now shaping future systems. Supported by tables and diagrams that clarify waste categories, regulatory differences, operational workflows, and technological advances, the chapter aims to offer readers not just information, but perspective. Ultimately, it seeks to help researchers, policymakers, and healthcare professionals understand what was learned, what must be improved, and how safer and more sustainable BMW can be built in a world changed by COVID-19.

5.2 Background

To manage biomedical waste effectively, we first have to understand what exactly we are dealing with. The most critical distinction lies between hazardous biomedical waste and nonhazardous general waste. Hazardous biomedical waste includes infectious materials contaminated with blood or pathogens, "sharps" like needles and scalpels, and pathological waste such as human tissues or organs. It also encompasses pharmaceutical waste (like expired drugs or cytotoxic agents), chemical waste from labs (reagents and heavy metals), and radioactive materials (Liu et al., 2015; Patil & Shekdar, 2001). On the other hand, general healthcare waste is essentially no different from household trash, think packaging, paper, and food scraps, and carries no specialized handling risks (Manekar et al., 2022).

The scale of this waste varies significantly by region. High-income countries typically generate about 0.5 kg of hazardous medical waste per hospital bed each day, while low-income countries produce around 0.2 kg per bed (Faisal et al., 2021). However, that lower figure in developing nations often masks the truth, because segregation is not always perfect, and hazardous waste is frequently underestimated. This makes proper segregation the absolute cornerstone of the system. By sorting waste correctly at the source, we ensure that only the truly dangerous materials undergo expensive, specialized treatment, while the remainder safely enter the standard municipal waste stream. To illustrate how this works in practice, Table 5.1 breaks down the major categories and their handling methods. This table uses India's color-coded system

Table 5.1 Color-coded categories of biomedical waste and disposal methods (India's BMW Rules, 2016).

Category (color code)	Types of waste	Primary treatment/disposal
Yellow	*Infectious waste*: items contaminated with blood or body fluids (e.g., soiled dressings, bandages, swabs); *Pathological waste*: human tissues, organs, body parts; *Pharmaceutical waste*: expired or discarded medicines; *Chemical waste*: discarded disinfectants, contaminated lab reagents; *Microbiological waste*: laboratory cultures, stocks Shammi et al. (2021); Wang et al. (2024).	Typically, **incineration** at high-temperature; alternatives include chemical disinfection or autoclaving for certain subtypes. Anatomical waste is often incinerated or deeply buried Wang et al. (2024). Residues (ash) are landfilled.
Red	*Contaminated recyclable waste*: primarily plastic items such as IV tubing, catheters, syringes without needles, urine bags, gloves, etc., that are contaminated with infectious materials, Celitron Medical Technologies (2019).	**Autoclave or microwave treatment** to sterilize, followed by **shredding**. Sterilized shredded plastic can be sent for recycling if facilities exist, Bagwan (2023). Nonrecyclable treated waste is landfilled.
White (Translucent)	*Sharps waste*: needles, syringes with attached needles, scalpels, blades, glass slides, any items that can pierce (whether contaminated or not), Ilyas et al. (2020).	**Autoclave/microwave plus shredding or encapsulation**. Sharps are collected in puncture-proof containers; after disinfection, they are shredded or encapsulated in cement and then landfilled. Metal sharps may be sent for metal recycling after treatment.
Blue	*Glass waste*: discarded medicine vials, ampoules, broken glassware; sometimes metallic implants, Bhar et al. (2022).	**Autoclaving** or chemical disinfection followed by **recycling or safe disposal**. Glass that has been sterilized can be crushed and recycled. If recycling is not feasible, treated glass is landfilled.

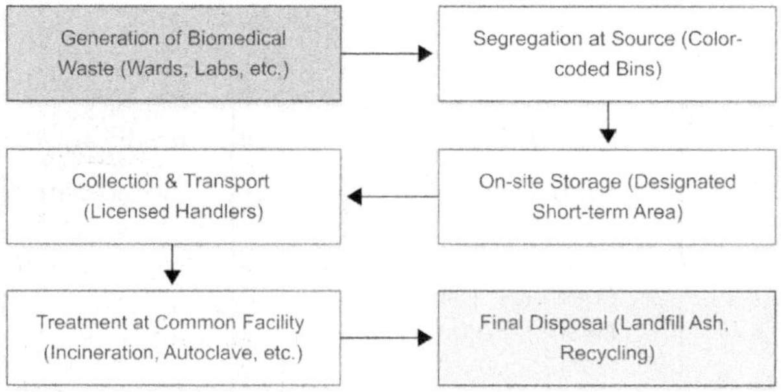

FIGURE 5.1

Standard lifecycle of medical waste, which begins the moment it is generated during patient care or diagnostics. At the source, it is immediately sorted into color-coded containers to prevent cross-contamination. After a short period of temporary storage within the hospital, it is usually collected by municipal services or external specialists and hauled to a treatment facility. There, the waste is neutralized through incineration, autoclaving (steam sterilization), or shredding. The final goal is safe disposal: turning hazardous materials into inert ash or stabilized waste for sanitary landfills, or even reclaiming recycled materials to re-enter the industrial supply chain.

mandated by the BMW Rules of 2016 as a primary example, though most international systems follow similar WHO-aligned color schemes and best practices.

Each type of waste is carefully managed to reduce its risks. Hospitals enforce segregation at the source by requiring healthcare staff to place waste into color-coded bins right where it is generated. Fig. 5.1 shows a typical hospital BMW workflow, covering every step from waste generation to final disposal. This multistage process includes sorting waste into the correct containers, temporary storage within the facility, collection by authorized personnel, treatment at specialized facilities, and the safe disposal of any remaining residues.

Before the arrival of COVID-19, global biomedical waste systems were on a path of steady improvement, yet significant gaps remained a reality. The consequences of unsafe disposal were already well-documented as a major public health hazard. For instance, the WHO reported that in 2010, the reuse or poor disposal of contaminated syringes led to approximately 33,800 new HIV infections and 1.7 million hepatitis B infections worldwide (World Health Organization 2022; Yang et al., 2021). In developing countries, the stakes were even higher, with an estimated 25% of the infectious disease burden linked directly to mismanagement, such as needle-stick injuries or the contamination of local water supplies (Ilyas et al., 2020). Consequently, there was a clear, preexisting global consensus on the need for stricter regulations and better infrastructure, like secure landfills and high-capacity incinerators (Ilyas et al., 2020). Many nations had already begun adopting rules that mirrored the WHO's "Safe management of healthcare waste" guidelines and the technical standards of the Basel Convention (Um et al., 2023).

The COVID-19 pandemic placed extraordinary strain on BMW systems, exposing just how unprepared many of them were for a crisis of this scale. Across the world, the volume of biomedical waste rose sharply and almost overnight. The routine use of single-use PPE, such as masks, gloves, gowns, and face shields, became a daily necessity rather than an exception (Baniak et al., 2025). At the same time, mass testing programs generated millions of swabs, vials, and diagnostic kits, while hospitals and quarantine facilities produced growing amounts of infectious waste as they cared for critically ill patients. As vaccination drives gathered momentum, another layer of waste was added in the form of syringes, needles, and vaccine packaging. Together, these overlapping demands pushed existing systems far beyond their intended capacity, bringing long-standing gaps in infrastructure, logistics, and oversight into sharp focus, issues that are examined in greater detail in the section that follows.

Globally, the numbers are staggering. A WHO analysis reported that between March 2020 and November 2021, approximately 87,000 tons of PPE were procured through a UN supply initiative, and nearly all of it is expected to have become waste (World Health Organization 2022). Similarly, over 140 million COVID-19 test kits were shipped, with an estimated 2600 tons of plastic waste (and 731,000 L of chemical waste) resulting from their use (Rankin et al., 2024). The worldwide vaccination drive administered about 8 billion doses, which translates to 144,000 tons of waste in the form of syringes, needles, and safety boxes for sharps disposal (Kim et al., 2024). These figures reflect just the *medical* waste; on top of this, billions of disposable masks used by the general public have contributed to municipal solid waste and litter. In many places, used masks and gloves began appearing on streets, beaches, and water bodies, highlighting lapses in proper disposal (Mijakoski et al., 2020; Platon et al., 2020). The WHO and national health authorities warned that improper dumping of PPE could facilitate virus spread (though evidence of fomite transmission was limited) and pose direct risks to waste workers and the environment (Kankanamge et al., 2020; Mejjad et al., 2021).

Regional experiences varied, but South Asia exemplified the challenges faced by developing countries. India and Bangladesh, with large populations and developing waste infrastructure, saw dramatic increases in medical waste. In India, the BMW generation rose sharply during the COVID-19 pandemic. The country produced about *710 tons per day* of biomedical waste, and an additional ~*126 tons/day* were attributable to COVID-19 during the first wave (around 20% increase) (Rajak et al., 2022). As the pandemic progressed, the waste kept mounting: By May 2021 (during the second wave), India's COVID-related biomedical waste peaked at ~203 tons per day, up from a peak of ~183 tons in 2020 (Capoor & Parida, 2021). Over the year leading up to May 2021, India generated a cumulative 45,954 tons of COVID-19 waste (tracked via a national reporting app) (Jindal & Sar, 2023). The sudden burden stressed treatment facilities, incinerators, and autoclaves were operating at capacity, and some cities struggled to cope with the volumes (Hossain et al., 2020; Kudli et al., 2021).

FIGURE 5.2

Improper disposal of COVID-19 healthcare waste: (A) Biohazard bags containing COVID-19 waste (PPE such as masks, gloves, head covers) thrown outside a dedicated COVID hospital; (B) Piles of COVID-related waste mixed with general garbage near a hospital. Such scenes were common during the pandemic surge in many South Asian cities (as per various media reports), underscoring lapses in segregation and safe disposal. Waste left in the open not only risks spreading infection but also contaminates the environment (note the proximity to public areas). These challenges highlight the urgent need for strengthened protocols and infrastructure for biomedical waste management.

Bangladesh faced a similarly dire situation. Even before COVID-19, the capital Dhaka produced an estimated 206 tons of medical waste per day on average (Ranjan et al., 2020), much of which was inadequately treated. During the pandemic, this volume further increased in tandem with infection waves (Mondal et al., 2022). A study in Dhaka found that in just 1 month of the early pandemic (March April 2020), over 14,000 tons of single-use plastic waste were generated from COVID-related protective gear and supplies: ~5877 tons of gloves, 1592 tons of masks, 5796 tons of plastic shopping bags (from heightened home-delivery and medical supply packaging), and hundreds of tons of sanitizer bottles (Mohammad et al., 2020). This *"voluminous amount of waste"* was poorly managed, creating environmental and public health hazards (Kothari et al., 2021; Kulkarni & Anantharama, 2020) (Fig. 5.2).

As these images suggest, the sheer volume of waste left many facilities in South Asia completely overwhelmed. In Bangladesh, reports indicated that the majority of hospitals, crippled by negligence or a lack of capacity, abandoned disinfection and segregation protocols entirely (Khosla et al., 2022; Singh et al., 2024). This pushed the burden onto municipal cleaners and informal waste pickers who, lacking both PPE and training, faced immense occupational risks while handling infectious materials (Khosla et al., 2022). Without designated disposal sites, this waste frequently ended up in open dumps or local waterways, creating significant downstream health risks for the broader community (Gupta et al., 2023). While India had a more established framework, similar cracks appeared in smaller towns and rural districts where hospitals were not yet integrated into centralized treatment networks (Singh et al., 2024).

Table 5.2 Examples of increases in medical waste generation during COVID-19.

Location	Pre-COVID waste generation (approx)	COVID-19 peak/additional waste	Increase factor/notes
Wuhan, China	~40 tons/day (medical waste) Luhar et al. (2022)	~240 tons/day at peak Luhar et al. (2022)	A six-fold increase (during the early 2020 outbreak). Existing incineration capacity (49 tons/day) was overwhelmed Luhar et al. (2022)
Ahmedabad, India	0.55 0.60 tons/day (medical waste), Liang et al. (2021)	~1.0 ton/day (during initial lockdown), Liang et al. (2021)	A ~1.7-fold increase. Similar spikes strained waste handlers in many Indian cities.
Dhaka, Bangladesh	*Baseline not well segregated* (estimated ~200 tons/day, including general waste)	~206 tons/day (mid-2020), Ranjan et al. (2020); further increased with case surges	**High burden** on limited facilities. Large portions of this waste remained untreated, exacerbating pollution, Khosla et al. (2022)
Global PPE waste	Negligible (pre-2020 baseline for disposable masks/gloves in community)	~87,000 tons (PPE supplied Mar 2020 Nov 2021), World Health Organization (2022)	**Massive new waste stream**. In addition, test kits added ~2600 tons of plastic and 730,000 L of chemical waste; vaccine syringes added ~144,000 tons, Rankin et al. (2024)

The data in Table 5.2 further illustrate this crisis, showing multifold increases in waste volume that rendered existing systems nearly obsolete. As storage space vanished and collection schedules crumbled under lockdown pressures, even established treatment facilities were forced to operate beyond their safe limits. While some regions, such as Wuhan, China, managed to deploy mobile incinerators and build new plants at breakneck speed (Adelodun et al., 2021; Tripathi et al., 2020), resource-constrained nations often had to resort to dangerous stopgap measures like on-site burial or open burning (Al-Emam & Al-Yousfi, 2021; Akcam et al., 2023). Furthermore, sustainability took a backseat to safety; recycling programs were curtailed by as much as 46% in the United States due to fears of virus transmission through manual sorting (Abbasi & Sıcakyüz, 2024; Almulhim et al., 2021; Jebaranjitham et al., 2022).

Ultimately, COVID-19 created a "perfect storm" that pushed BMW to its breaking point. The experience in South Asia, in particular, proved that a public health emergency can rapidly evolve into an environmental disaster if segregation and treatment capacities are not robust enough to handle a surge (Gupta et al., 2023; Khosla et al., 2022). However, as devastating as this period was, it also catalyzed reform. This crisis sparked a wave of policy adaptations and technological investments across the globe that are now beginning to redefine how we handle medical waste in the postpandemic era.

5.3 Hospital-level practices and challenges during COVID-19

BMW at the hospital-level is the essential frontline in preventing infections. Healthcare facilities are responsible for segregating waste right where it is produced, securely storing it, managing its internal transport, and ensuring its safe transfer to licensed treatment centers or municipal disposal systems. The COVID-19 pandemic not only tested these protocols but also led to their widespread strengthening and adaptation worldwide. Standard hospital procedures rely on a carefully designed safety framework, including the strategic placement of color-coded bins throughout patient care areas and thorough training for both clinical staff and cleaning personnel to ensure proper waste segregation. Key safety measures include the use of puncture-resistant containers for sharps, daily internal waste collection, and regular disinfection of storage spaces. In many locations, large hospitals simplify waste handling by contracting authorized bio-waste services that manage final treatment and disposal in compliance with strict regulations.

In India, for instance, it is a legal requirement for every hospital to maintain a formal BMW plan and keep meticulous records of waste generation and handover (Das et al., 2021). However, the reality of implementation often clashes with the ideal, particularly under severe resource constraints. Research conducted in India before the pandemic highlighted a persistent gap between regulation and practice (Tripathi et al., 2020). Despite the presence of detailed rules, common failures included improper segregation, such as the dangerous mixing of medical and general waste, as well as overflowing bins and a lack of basic protective gear for sanitation teams (Goswami et al., 2021; Selvan Christyraj et al., 2021). These challenges were not unique to India; similar or even more acute conditions were frequently observed across hospitals throughout South Asia and Africa.

COVID-19 made proper waste segregation and infection control even more critical. Hospitals dealing with COVID patients had to treat all waste from isolation wards, intensive care units (ICUs), and testing labs as potentially infectious. New guidelines for hospital waste management during COVID-19 were quickly rolled out by health authorities. For example, in India, the Central Pollution Control Board (CPCB) issued specialized COVID-19 biomedical waste handling guidelines (first in March 2020, updated several times) (Bhar et al., 2022). Key measures included:

- *Dedicated color-coded bins in COVID-19 areas*: Ensure clearly labeled bins (yellow, red, etc.) are placed in COVID wards/ICUs for immediate segregation of waste (Ramteke & Sahu, 2020). All used disposable PPE, tubing, swabs, etc., from COVID patients should go into these bins.
- *Double-layer bags*: Waste from COVID isolation areas must be collected in a *double-layered yellow bag* to prevent leaks (Sawant & Kashte, 2024). This extra precaution addressed the higher infectious load and liquid content of some wastes (as per Indian Council of Medical Research guidelines) (Dehal et al., 2022).
- *Separate storage and labeling*: Hospitals were required to have a separate, dedicated *temporary storage area* for COVID waste, isolated from other waste streams (Behera, 2021; Kanwar et al., 2023). Bags and containers had to be labeled as "COVID-19 waste" to alert waste handlers (Khosla et al., 2022).
- *Timely collection and disposal*: COVID wards' waste should be disposed of *immediately or at least daily*. Many places mandated that COVID waste be removed within 24 h and sent for treatment without delay (Hantoko et al., 2021). Some hospitals increased the frequency of internal waste collection shifts.
- *Disinfection of bins and trolleys*: All bins, trolleys, and surfaces that came in contact with COVID waste were to be disinfected daily with 1% sodium hypochlorite or another effective disinfectant (Barua & Hossain, 2021). This was to protect cleaning staff and prevent surface contamination.
- *Record-keeping and reporting*: Hospitals had to maintain a separate log of the quantity of COVID-19 waste generated in isolation areas (Kuppusamy et al., 2022). In India, these figures were reported through an app to CPCB for monitoring (Ojha et al., 2022). Regular reporting helped authorities track trends and ensure waste was reaching treatment facilities.
- *72-h storage for home quarantine waste*: Recognizing that many COVID patients were isolating at home, guidelines advised that used masks, gloves, and other disposables from home quarantines be kept in a paper bag for a minimum of 72 h *before* disposal with domestic refuse (Hamid et al., 2013; Sujon et al., 2022). The 72-h period was based on evidence about the virus's survival time on surfaces and aimed to reduce the infectivity of the waste. Such waste, after 3 days, could be put in general trash (preferably after cutting or shredding items like masks to prevent reuse) (Barua & Hossain, 2021; Singh et al., 2022).

Though straightforward in theory, these measures were difficult to apply consistently across all hospitals. Facilities faced rising patient numbers while needing to enforce stricter waste management protocols. In well-equipped hospitals, infection control teams trained cleaning staff on updated standard operating procedures (*SOPs*), ensured sufficient supplies of bins and bags, and closely monitored adherence. However, many overburdened hospitals experienced lapses in implementation.

A glaring issue was the *safety of sanitary and waste management workers*. During the pandemic's peak, these workers faced a higher risk due to the infectious nature of waste. Ideally, all personnel handling COVID waste should wear appropriate PPE (gloves, masks, face shields, and gowns) and be trained in donning and doffing procedures to avoid self-contamination. In reality, frontline waste handlers (often low-paid hospital janitors or municipal workers) **did not always receive adequate PPE or training**, especially in developing countries (Gupta et al., 2023; Khosla et al., 2022). Tragically, reports emerged of waste workers contracting COVID-19, presumably via contact with contaminated materials. This underscored that the welfare of these often-forgotten workers is integral to an effective waste management system.

Another challenge was space: hospitals needed to allocate separate storage for COVID waste. Many small or rural hospitals lacked any dedicated waste storage room, so they had to improvise (e.g., cordoning off an area or using large containers with lids). Without proper storage, waste could pile up in wards or corridors, increasing infection risk.

Nonetheless, numerous **hospital-level innovations and good practices** were seen: Some hospitals set up on-site autoclaves to immediately treat COVID waste and reduce volume before off-site disposal. Others implemented *"just-in-time" waste collection*, increasing pickup frequency by waste services to prevent accumulation. Hospitals in several countries also tried to minimize waste generation by adopting reuse strategies when safe, for instance, disinfecting face shields for reuse or using washable gowns in place of disposables (when supply chains allowed). One study even explored reusing N95 respirators after decontamination; methods like vaporized hydrogen peroxide and UV irradiation were successfully used to extend the life of scarce N95 masks (Hamid et al., 2013). Although the reuse of single-use PPE was an emergency measure, it has prompted hospitals to consider more sustainable practices going forward.

To visualize how waste is segregated within a hospital during COVID times, Fig. 5.3 shows a schematic of the segregation process with the added emphasis on isolation ward waste.

Despite the challenges posed by the pandemic, a positive development at the hospital-level has been increased awareness of proper waste management. Many healthcare workers, who previously may have overlooked waste segregation, became more vigilant due to the heightened infection risks. Hospitals are expected to maintain some of these enhanced practices, such as stricter disinfection routines and improved staff training beyond the pandemic. However, lasting progress will depend on strong support from national policies and sufficient infrastructure.

5.4 National policy frameworks: Regulations and COVID era adjustments

National regulations provide the structural backbone for BMW, explicitly defining the duties of healthcare facilities, waste operators, and government oversight bodies. Over the last 20 years, most nations have established legal frameworks largely mirroring

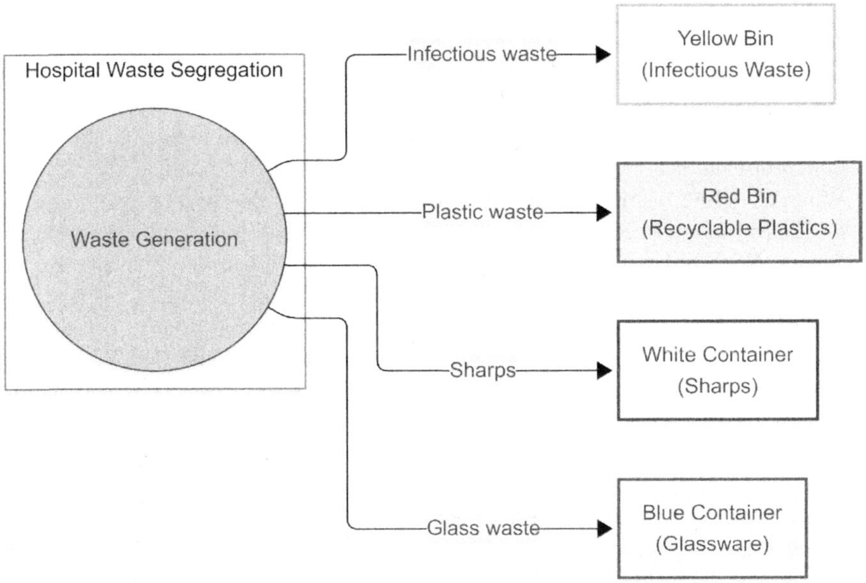

FIGURE 5.3

COVID-19 considerations: All waste generated is **segregated at source**. Infectious waste (especially from COVID-19 patients) is put in **yellow bags/bins**; contaminated plastics in **red bins** for disinfection and recycling; sharps in **white puncture-proof containers**; glass in **blue boxes**. In COVID-19 isolation areas, *all waste* is treated as infectious (thus often double-bagged in yellow containers). Sharps from COVID areas also go into their designated containers but are handled with extra precautions (indicated in pink). This segregation ensures that high-risk waste is set apart for specialized treatment methods like incineration or autoclaving, while lower-risk waste is handled through appropriate processes.

the WHO recommendations that standardize waste classification, segregation, transportation, and treatment technologies. Within South Asia, this regulatory landscape has matured significantly, though the rigor of enforcement varies.

- *India*: India stands as a primary example of a highly structured approach; its **BMW Rules, 2016 (amended in 2018),** provide an exhaustive legal roadmap that mandates color-coded segregation and requires every healthcare facility to pretreat infectious lab waste on-site before it is transferred for final disposal (World Health Organization 2022). Central to India's strategy is the use of **Common Biomedical Waste Treatment Facilities (CBWTFs)**, which are centralized plants equipped with incinerators, autoclaves, and shredders that service clusters of hospitals. By 2020, India had established roughly 200 of these authorized facilities, supported by thousands of smaller on-site units in areas where centralized access is lacking

(Jindal & Sar, 2023). Furthermore, the framework introduced significant environmental safeguards, such as the ban on chlorinated plastic bags and gloves to prevent toxic dioxin emissions during incineration (Wang et al., 2024). These rules are strictly enforced by **state pollution control boards**, and the accountability measures, including mandatory record-keeping and annual reporting, ensure that noncompliance can lead to severe penalties or the forced closure of the facility. When COVID-19 struck, India was quick to issue *supplementary guidelines under the 2016 rules*. The CPCB's COVID-19 waste guidelines (initially issued in March 2020 and periodically updated) essentially built upon the existing rules with COVID-specific details (Bhar et al., 2022; Gill et al., 2022). As discussed in the previous section, these included requirements like double bagging and labeling of COVID waste, separate records, and use of the CPCB's *COVID-19BWM* online application for daily waste reporting (Khosla et al., 2022). The fact that India already had a regulatory system in place helped in managing the surge. For instance, every state's Pollution Control Board was tasked to ensure that COVID hospitals were tied into the CBWTF network, and where needed, temporary incinerators or deep burial pits were approved under emergency provisions. The national rules were not fundamentally changed by COVID-19, but the pandemic did lead to tighter implementation. The crisis also prompted the government to consider more permanent changes: increasing the number of treatment facilities, updating pandemic waste handling protocols in the Disaster Management Act, and encouraging waste *minimization* strategies (like reuse of PPE).

- *Bangladesh*: Bangladesh introduced the *medical waste (management and processing) rules* in **2008**, one of the earlier South Asian nations to do so (Siddiqui et al., 2021). This regulation, under the Department of Environment, requires hospitals to segregate waste into six categories (with color codes: e.g., yellow for infectious, red for sharps, etc.), and to either treat waste on-site or hand it to city corporation-authorized waste handlers (Al-Omran et al., 2021). It mandated the formation of Waste Management Teams in each hospital and encouraged public-private partnerships for waste treatment. However, enforcement of the 2008 rules has been weak. By 2020, only a fraction of hospitals (mostly in major cities) had proper waste management systems. In Dhaka, a few NGOs and private companies (such as **PRISM Bangladesh**) work under contract to collect and treat medical waste from some hospitals (Bizuneh et al., 2024; Nayan et al., 2021). Their capacity covers only about half of the total medical waste generated, leaving the rest to be mixed with municipal garbage or dumped (Behera, 2021). The COVID-19 pandemic caught Bangladesh's system unprepared initially; there were no specific COVID waste guidelines, and hospitals largely followed their own ad-hoc practices. By mid-2020, the government, with input from NGOs, issued some **advisories** (for example, urging double bagging of COVID waste, use of PPE by cleaners, and chlorine spray on waste), but these were not legally binding. Realizing the gap, policymakers in Bangladesh are now considering updating the 2008 rules. The post-COVID agenda includes establishing more centralized treatment facilities, expanding coverage beyond major cities, and integrating pandemic waste strategies into hospital accreditation.

- *Pakistan:* Pakistan's federal government implemented *Hospital Waste Management Rules* in **2005** (Voudrias, 2024), which were among the first in the region. These rules categorize hospital waste into infectious, toxic, sharps, etc., and lay out requirements for handling and disposal (much of it in line with WHO guidance). Hospitals are expected to have waste management plans and committees. In practice, enforcement is left to provincial authorities and has been inconsistent. Some large hospitals in Pakistan have incinerators on-site, while others rely on municipal services that often dump waste in open sites. During COVID-19, Pakistan issued *SOPs for COVID-19 waste management* as an annex to its National Action Plan for COVID-19 (Janik-Karpinska et al., 2023; Manupati et al., 2021). The SOPs largely reinforced existing rules: Treat COVID waste as infectious clinical waste, ensure waste worker protection, etc. There was a push to use lime pits or incineration for the disposal of PPE and other COVID waste. But similar to Bangladesh, the absence of a robust waste collection/treatment infrastructure hampered proper implementation in many areas. Moving forward, Pakistan is exploring public-private partnerships and international assistance to upgrade its medical waste systems, including plans to install modern autoclaves and shredding units in major hospitals. Table 5.3 compares key elements of the biomedical waste regulations and COVID era measures in India, Bangladesh, and Pakistan (as representative South Asian countries). It also notes relevant international guidelines that influence national policies.

India's existing BMW system demonstrated considerable resilience and was effectively utilized during the pandemic. In contrast, neighboring countries like Bangladesh and Pakistan encountered greater challenges due to enforcement difficulties and gaps in infrastructure (Maalouf et al., 2022; Shbaklo et al., 2021). Still, the pandemic acted as a powerful wake-up call for all. National governments now recognize that managing biomedical waste is not merely an environmental concern but a vital component of public health and emergency preparedness.

At the national level, both in South Asia and globally, several important policy changes have begun to take shape. Legal amendments are being drafted to ensure pandemic scenarios are no longer handled as "exceptions." For instance, India is moving to formalize temporary measures such as mandatory digital tracking into permanent law. Simultaneously, we are seeing a massive drive for infrastructure investment. Governments are securing international funding, such as Pakistan's collaboration with the Asian Development Bank, to build treatment capacity in underserved regions through new incinerators and autoclaves (Ranjan et al., 2020).

The "human element" of the system is also receiving excessive attention. National health ministries, such as those in Bangladesh, are partnering with NGOs to launch comprehensive training programs for hospital staff and waste handlers, prioritizing the correct use of PPE and rigorous segregation. Furthermore, the crisis forced a collapse of traditional silos. The necessity of the pandemic response fostered rare interagency

Table 5.3 National biomedical waste management frameworks and COVID-19 measures.

Country	BMW legislation and key features	COVID-19 specific measures (during pandemic)
India	**BMW Rules, 2016** (Ministry of Environment) Mandates segregation into four colored categories Capoor & Parida (2021), use of CBWTFs; strict record-keeping and annual reporting; bans chlorinated plastics in incineration. Enforcement by pollution control boards, World Health Organization (2022). Amendments in 2018 improved barcoding and monitoring.	**CPCB COVID-19 Guidelines (2020)** required double bagging of COVID waste, labeling as "COVID-19 waste," separate storage; launched a **mobile app for daily waste tracking**Ojha et al. (2022); mandated PPE for waste handlers and 72-h predisposal storage for home quarantine waste Barua & Hossain (2021). These guidelines were revised ~4 times in 2020 21 to adapt to evolving needs.
Bangladesh	**Medical Waste (Management and Processing) Rules, 2008** introduced a six-color segregation system, hospital waste management plans, and authorized city corporations to handle BMW (Al-Omran et al., 2021). Few centralized treatment facilities; mostly dependent on municipal collection or NGOs (e.g., PRISM) in cities. Enforcement has been limited (Behera, 2021).	**Interim COVID-19 waste advisories (2020)** Government circulars encouraged hospitals to segregate and double-bag COVID waste, disinfect waste with bleach, and ensure safe disposal (incineration or deep burial if possible). Due to limited capacity, much COVID waste was managed by ad-hoc means. Post-2020, policy focus is on upgrading facilities and possibly updating the 2008 Rules to incorporate pandemic preparedness.

(continued on next page)

Pakistan	**Hospital Waste Management Rules, 2005**, define categories (infectious, sharps, pharmaceutical, etc.) and assign responsibility to hospital management (Voudrias, 2024). Emphasize incineration or deep burial for hazardous waste, and disinfection of waste. Provincial regulations complement federal rules in some regions.	**COVID-19 waste SOPs (2020)** Issued as part of Pakistan's COVID response plan, Janik-Karpinska et al. (2023). Advised treating all patient waste from COVID isolation as infectious; using dedicated bins and bags; immediate transport to incinerators. Some cities deployed quick fixes (e.g., digging lime-lined pits for PPE disposal in remote areas). Efforts were made to provide PPE to sanitary staff. Ongoing plans to invest in modern autoclaves and improve compliance.
International	**WHO guidelines:** "Safe management of wastes from healthcare activities" (the "Blue Book") provides global best practices. **Basel Convention:** Technical guidelines (2011) on biomedical and healthcare waste management offer standards for environmentally sound disposal (Um et al., 2023). **Other**: World Bank and UNEP programs supporting waste infrastructure.	**WHO Interim Guidance (2020):** WHO published notes on COVID-19 waste urging strong segregation, use of impermeable bags, and high-temperature incineration or autoclaving for infectious waste (Zarocostas, 2022). Also recommended that *waste from COVID home care* be securely bagged for municipal collection (Zarocostas, 2022). In 2022, the WHO called for **"integrating better, safer, and greener practices"** post-COVID, including nonburning technologies and reusable PPE (Maalouf et al., 2022).

coordination, bringing health officials, city corporations, and environmental regulators together into unified task forces. This collaborative "build back better" ethos is now being codified into joint protocols to ensure capacity buffers and emergency plans are ready for any future surge.

On a global scale, international organizations continue to provide the vital scaffolding for these national efforts. The WHO remains the primary architect of these standards. Its "Blue Book" provides the definitive roadmap for safe healthcare waste management, advocating for a clear hierarchy that prioritizes waste minimization and segregation (Lyu et al., 2025; Singh et al., 2020). Crucially, the WHO's current guidance pushes for a transition toward nonincineration treatments like autoclaving to mitigate harmful emissions, while reinforcing the non-negotiable need to protect waste workers and the surrounding community from exposure (Bank, 2020; Hackett, 1989). During COVID-19, the *WHO released interim guidance notes specifically on waste management in the context of the pandemic*. Key points from the WHO and other international agencies included (Zarocostas, 2022).

- *Maintain essential waste services*: Even during lockdowns, waste collection from healthcare facilities and communities should continue regularly as an essential public service (Al-Emam & Al-Yousfi, 2021). Any disruption can lead to the accumulation of infectious waste.
- *Safe handling of community PPE waste*: Used masks and gloves from households should be considered potentially infectious. WHO recommended disposing of them in strong bags, tying them securely, and having municipal waste services collect and treat them, rather than leaving them in open litter (UNEP 2020).
- *No special protocols beyond standard IPC needed for COVID in hospitals*: Interestingly, WHO and the US CDC noted that **existing infection prevention and control (IPC) procedures for medical waste are effective against SARS-CoV-2**, so long as they are rigorously applied (Ugandar et al., 2023). This means hospitals that properly segregate and autoclave/incinerate infectious waste do not need additional novel methods; just strict compliance. CDC's guidance was that medical waste from COVID patients can be managed **"in accordance with routine procedures"** as for any regulated medical waste (Ugandar et al., 2023). This stance was important to prevent overreaction and unsafe experimental practices. However, the challenge was that in many settings, the "routine procedures" themselves were not fully in place or followed.
- *Environmental sustainability concerns*: In early 2022, WHO sounded an alarm on the *mountains of COVID-19-related waste* and urged healthcare systems to incorporate sustainability in their recovery plans (Mohamed et al., 2023; Singh et al., 2022). The WHO report (Feb 2022) highlighted not only the numbers (like 87,000 tons PPE, etc., discussed earlier) but also the strain this put on waste management systems worldwide (Singh et al., 2022). It recommended a series of actions, such as **using eco-friendly and reusable PPE, investing in nonburning technologies for waste treatment (autoclaves, microwave, etc.), and boosting the recycling sector to handle plastic waste**

(Maalouf et al., 2022). Essentially, the global guidance is shifting toward a *greener approach,* acknowledging that while infection control is paramount, we should avoid simply incinerating everything and instead seek ways to reduce and recycle waste safely.

Beyond WHO, other international frameworks relevant to biomedical waste include:

- *The Basel Convention (UN)*: This is a global treaty (effective 1992) on the control of transboundary movements of hazardous wastes and their disposal. *Infectious medical waste* is classified as hazardous under Basel (Category B2, "wastes from healthcare activities"), and its international transport is tightly regulated. More practically for domestic management, the Basel Convention published detailed **technical guidelines** on biomedical and healthcare waste (most recently updated in 2003) (Um et al., 2023). These guidelines cover waste minimization, segregation, storage, transportation, treatment, and final disposal options, aligning with the environmentally sound management (ESM) principles. They also endorse technologies like autoclaving, microwave treatment, and *coincineration in cement kilns* as viable options for biomedical waste disposal (Mohamed et al., 2024). During COVID-19, the Basel Convention Secretariat collected and disseminated information on how different countries were coping with pandemic waste, and encouraged parties to follow the technical guidelines even under emergencies (Shekoohiyan et al., 2021).
- *United Nations Environment Program (UNEP) and International Agencies*: The UNEP and other agencies (like UNICEF, UNDP) have been involved in biomedical waste issues, particularly in developing countries. During the pandemic, UNEP noted that waste management is an essential service and published a *Basel Convention fact sheet* on healthcare waste specific to COVID-19 (Budati et al., 2024). The World Bank also provided funding and guidance to some countries for strengthening waste management as part of COVID-19 emergency support (for example, procuring autoclaves or safety boxes for syringes in vaccine campaigns). There is also the **UN Environment Management Group (EMG),** which released guidance on COVID-19 waste, emphasizing that if not managed properly, infected waste can lead to secondary disease outbreaks and environmental pollution (Najafighodousi et al., 2024).

International professional bodies (like the International Solid Waste Association, ISWA) issued advisories too, often echoing WHO's points. There was consensus globally that **the best practices for COVID-19 waste are essentially the best practices for any infectious waste**, implemented universally. The difference was largely one of scale and ensuring continuity of operations under pandemic conditions.

One important global aspect is the **sharing of knowledge and innovations**. Over 2020 2021, numerous webinars, publications, and guidance documents were shared through international channels so that countries could learn rapidly from each other. For instance, experiences from Wuhan's handling of waste (scaling up incineration and deploying mobile treatment units) were documented and made available as a reference for other cities (Ali et al., 2022; Amin et al., 2024; Barua & Hossain, 2021). This global diffusion of knowledge was valuable for countries that were later hit by waves of infections.

The global regulatory environment provided a supportive framework during COVID-19, but it also gained new prominence. The pandemic has inserted waste management firmly into the global health security agenda. At forums like the World Health Assembly and UN climate and environment meetings, there are now calls to invest in **"climate-smart" healthcare waste management as part of resilient health systems** (Mohamed et al., 2023). The idea is that as the world prepares for future health emergencies, ensuring sustainable and safe waste handling capacity is a key component benefiting both health and the environment.

5.5 Innovations in biomedical waste treatment, tracking, and disposal

One of the undeniable silver linings of the COVID-19 crisis has been the rapid acceleration of innovation across the biomedical waste sector. Faced with a logistical nightmare, stakeholders ranging from hospital administrators to tech developers and researchers were forced to deploy solutions that, under normal circumstances, might have taken a decade to reach the field. This section explores the notable breakthroughs in tracking and data management that have defined the post-COVID era.

1. *Digital tracking and data management*: The pandemic proved that we can no longer manage waste "in the dark," as the need for real-time visibility into waste generation led to the rapid adoption of digital logging tools. A standout example is the smartphone app launched by India's CPCB in May 2020. By requiring facilities to log daily outputs, authorities could finally aggregate data to identify surging hotspots and allocate disposal resources before systems became overwhelmed (Ojha et al., 2022). Beyond simple apps, we are seeing the rise of a true Internet of Things (IoT) ecosystem within hospital walls. IoT-enabled "smart bins" equipped with sensors can now monitor fill levels in real-time, alerting collection teams only when a pickup is actually necessary, which optimizes logistics and reduces unnecessary exposure (Idrees et al., 2022). Furthermore, the use of RFID tags and QR codes has moved from retail into the hazardous waste ward. By tagging waste at the point of origin, facilities can track a specific batch through every stage of transport to the final treatment site (Srivastava et al., 2024). This level of digitization is not just about efficiency; it serves as a critical safeguard against the illegal "leakage" of waste. During the early days of the pandemic, reports surfaced of scavenged PPE being illicitly resold (Ali & Parvin, 2022). A transparent, digital paper trail makes such pilferage of dangerous items like used syringes far more difficult. Looking ahead, the logical next step is the transition from "optional innovation" to "mandated standard" (Ishaq et al., 2025). We expect national regulations

to increasingly require these tracking systems, as this data-driven approach allows us to size future treatment capacity-based on actual generation trends and provides an early warning system if a facility's disposal patterns become irregular.

2. *Novel treatment technologies*: Traditional biomedical waste treatment relies heavily on **incineration** (for infectious and sharp wastes) and **autoclaving** (steam sterilization, often followed by shredding, for many types of waste). During COVID-19, both of these were used extensively. However, concerns about air pollution from incinerators (emitting pollutants like dioxins, mercury, etc.) and the sheer volume of waste have spurred interest in alternative technologies that are cleaner or can handle large loads:
   - *Plasma pyrolysis*: This is an emerging technology where a plasma torch is used to thermally degrade waste at extremely high temperatures (>1200°C) in an oxygen-starved environment. It can break down even complex wastes (including plastics and pathogens) into syngas and a vitrified inert residue, with minimal emissions if properly controlled. Plasma units have been piloted for medical waste in some countries. The pandemic's waste burden renewed interest in plasma as a decentralized solution (since units can be small-scale). Some Indian institutes, for instance, have developed plasma pyrolysis units to treat PPE and infectious waste on-site (Capoor & Parida, 2021). The advantage is the *drastic volume reduction* and elimination of potentially infectious agents without smoke; the challenge is high energy consumption and cost.
   - *Microwave irradiation*: Already recognized by WHO as an alternative, microwave-based treatment involves shredding waste and then heating it with microwaves to disinfect (usually with some moisture present to create steam). This method can be as effective as autoclaving and has been scaled up in some places to cope with pandemic waste. For example, a few cities in China and Europe brought in mobile microwave treatment trucks during COVID-19, which could be driven to hospitals to process waste on-site, reducing transportation needs (Kashyap et al., 2020).
   - *Chemical disinfection (advanced)*: Chemical treatment (using chlorine compounds, ozone, etc.) saw innovations, too. **Ozone disinfection chambers** for medical waste were explored. Ozone is a powerful oxidizer that can kill pathogens. Some hospitals used large ozone boxes to disinfect PPE for either safe disposal or reuse (Chand et al., 2021). There is also research on **enzymatic treatments** for certain medical wastes, though not yet widely applied.
   - *Superheated steam and novel autoclaves*: Conventional autoclaves operate at ~121°C. Newer systems using superheated steam (up to ~150–180°C) claim faster and more thorough sterilization, and some integrate grinding/shredding so that output is unrecognizable confetti (Hussein et al., 2022). These integrated sterilizer-shredder (ISS) systems became popular for on-site hospital use during the pandemic, as they can reduce waste volume by ~80% and render it noninfectious, allowing hospitals to simply send the residue to

municipal landfills instead of needing special incineration (Bank, 2020; Ranjan et al., 2020).

- *Cement kiln coprocessing*: A pragmatic innovation, especially in developing countries, has been to utilize existing cement kilns to coincinerate biomedical waste. Cement kilns operate at very high temperatures (~1400°C) and can effectively destroy organic waste, with the added benefit that the ash gets embedded in the clinker (so no residue to landfill). During COVID-19, some places (e.g., in Thailand and India) directed excess medical waste to cement plants as an emergency measure. This required careful handling and limited quantities, but it opened eyes to an existing infrastructure that can be tapped in waste crises (Mohamed et al., 2024).

3. *Waste volume reduction and recycling innovations*: With plastics comprising a large portion of COVID-related waste (masks, gowns, and test kits), finding ways to recover value or at least reduce harm from this waste has been a focus of innovation:
   - *Recycling of PPE plastics*: Researchers have developed methods to recycle polypropylene (the plastic in surgical masks and PPE coveralls) once it is decontaminated. One study demonstrated that shredded disposable masks can be added to asphalt for road construction, reinforcing while diverting waste from landfills (Soares et al., 2013). Approximately 3 million masks could be reused per kilometer of a two-lane road, according to that study (Aliyi et al., 2024). Another approach is mechanical or chemical recycling, e.g., Australian scientists shredded masks and mixed them with recycled concrete to create construction bricks and materials (Ojha et al., 2022; World Health Organization 2024). While not yet mainstream, these innovations show promise for handling the *deluge of PPE waste* in a circular way.
   - *Waste-to-energy and fuel from plastics*: Pyrolysis (in a controlled reactor, not necessarily plasma) can convert medical plastic waste into synthetic oil or gas. There are pilot projects that have used pyrolysis on discarded PPE to create fuel for industrial use (Haque et al., 2021). Similarly, some incinerators were coupled with energy recovery systems (boilers to capture heat). During the pandemic, such waste-to-energy helped in places like Wuhan to manage waste while also generating steam/electricity for use (Česonienė et al., 2022; Ganguly & Chakraborty, 2021).
   - *Reusable and eco-friendly materials*: Innovation goes beyond just managing waste at the end of its lifecycle; it also involves redesigning products to minimize waste from the start. In response to these challenges, the global health community has begun to look more seriously at reusable alternatives to conventional disposable PPE. Washable gowns and cloth masks, particularly for use in noncritical and low-risk settings, are increasingly seen as practical options that can substantially cut down waste without compromising safety. Some manufacturers have gone a step further, developing autoclavable face masks fitted with replaceable filters, allowing the same mask body to be safely reused multiple times. Reflecting this shift in thinking, the **WHO**

has explicitly encouraged investment in reusable and longer-lasting medical materials, framing them not only as an emergency response to pandemic-related shortages but as a longer-term strategy for building more sustainable healthcare systems (World Health Organization 2022). We also saw 3D printing used to produce face shield frames and mask components that were reusable, an innovative approach to alleviate supply issues and reduce disposables. Moving forward, hospitals are likely to stockpile more reusable gear and sterilization units for them.

4. *Automation and safety innovations*: As concerns about staff safety intensified, some hospitals began exploring automation to minimize direct human contact with potentially infectious waste. In certain settings, simple robotic solutions were tested, such as small, mobile carts that transported sealed waste bags from COVID wards to temporary storage areas, thereby reducing repeated exposure for housekeeping and nursing staff. The idea of using drones for medical waste transport, particularly for samples or very small waste loads, also emerged in discussions. Although promising in theory, these approaches are currently limited by regulatory, safety, and airspace management challenges, hindering their widespread adoption. Alongside automation, attention shifted to surface-level safety innovations. Antimicrobial coatings on waste bins and collection bags were introduced to mitigate the risk of contamination during handling. In India, for instance, a manufacturer began producing biomedical waste bags embedded with antimicrobial layers designed to inactivate viruses on contact, providing an additional layer of protection for waste handlers. Importantly, these developments reflect a broader shift in thinking. Innovation in BMW is no longer confined to end-stage treatment and disposal. Increasingly, it involves re-evaluating workflows, materials, and procurement practices to reduce waste generation at the source. Some healthcare systems have applied lean management principles and data-driven monitoring to streamline processes, minimize unnecessary disposables, and enhance overall efficiency, signaling a move toward more proactive, system-level solutions rather than reactive fixes.
5. *Process improvements and analytics*: By monitoring waste production through new tracking systems, hospitals can better control procurement, avoiding excess supplies that contribute to waste, and optimize waste handling logistics. The introduction of a "waste management dashboard" enables administrators to track, almost in real-time, the volume of waste generated by each department, associated disposal costs, and identify operational bottlenecks. This data-driven approach fosters more informed decision-making and greater accountability at the facility level (Table 5.4).

While these innovations show great promise, most remain in pilot or early adoption stages. To transition from experimental use to widespread implementation, robust government support and international scientific collaboration are crucial. Practical investments are needed to equip hospitals with advanced technologies like smart

Table 5.4 Innovations in biomedical waste management emerging in the post-COVID era.

Innovation	Description and example	Benefits
Digital tracking (apps and IoT)	Use of mobile apps and IoT sensors to monitor waste generation and movement. *E.g.,* CPCB's COVID waste tracking app logs daily waste from hospitals, Ojha et al. (2022); Smart bins with fill-level sensors and QR-coded bags for traceability, Srivastava et al. (2024); Sharma et al. (2020).	Improves compliance and planning. Real-time data helps allocate resources and ensures waste is collected and treated promptly. Reduces illegal dumping or reuse through end-to-end traceability.
Advanced sterilization tech	*Plasma pyrolysis units* achieving high-temp destruction of waste, Capoor and Parida (2021); *Integrated autoclave-shredder systems* for on-site treatment; *Ozone-based disinfectors* for PPE.	Provides alternatives to incineration with potentially lower emissions. On-site treatment reduces transport risks and volume of waste (by shredding). Can be scaled to areas lacking big facilities.
Reuse and reprocessing of PPE	Decontamination and reuse of certain PPE items. *E.g.,* UV or vapor hydrogen peroxide systems to reprocess N95 masks (used by some hospitals in 2020); reusable gowns and face shields introduced to replace disposable ones, World Health Organization (2022).	Dramatically cuts down waste generation and alleviates supply shortages. If done safely, it maintains protection while reducing cost and environmental burden.

(*continued on next page*)

Recycling and valorization of waste	Converting biomedical plastic waste into useful products. *E.g,* Shredded masks and gloves added to asphalt for road construction, Soares et al. (2013); pyrolysis of PPE to produce fuel oil, Haque et al. (2021). Additionally, recycling sterilized glass vials and plastic containers.	Recovers value from waste and reduces landfill load. Can stimulate a recycling market for medical plastics. Also mitigates pollution by ensuring these materials are contained in new products rather than littering ecosystems.
Waste worker safety innovations	Tools and protocols to protect waste handlers: *E.g,* touch-free waste trolleys or robotic waste carts; antimicrobial waste bags that reduce surface pathogen survival; dedicated PPE kits and vaccination for sanitation staff.	Reduces infection risk for a vulnerable group of workers. By making waste handling safer, it ensures the system keeps running even during health crises. Also elevates the dignity and recognition of sanitation workers as essential health protectors.

autoclaves and to establish specialized recycling facilities capable of safely processing medical-grade plastics. Simultaneously, innovation must be accompanied by regulation. There is an urgent need for standardized guidelines to govern emerging technologies, including certification protocols for plasma pyrolysis emissions and safety standards to ensure recycled medical materials are thoroughly decontaminated for reuse. The COVID-19 pandemic served as a powerful catalyst, compressing what would have been a decade of gradual progress into just 2 years of rapid, necessity-driven innovation. Waste management institutions and contractors that had previously resisted change quickly adopted new methods, while regulators became more flexible in approving alternative treatment options. The silver lining is that we now possess a much broader toolkit for BMW than we did in 2019. Our next collective challenge is to stop treating these innovations as emergency measures and start integrating them into the permanent fabric of routine practice and global regulations. The positive outcome is that we now have a much broader array of tools for BMW than we did before the pandemic. The next challenge is to move beyond treating these innovations as temporary emergency measures and fully integrate them into standard practice and global regulatory frameworks.

5.6 Policy modifications during and after COVID-19

As discussed throughout this chapter, numerous **policy modifications** were implemented in reaction to the COVID-19 crisis, and many of these have lasting implications. In this section, we synthesize the key changes that occurred at different levels of hospital protocols, national policies, and global guidelines and consider how they shape the postpandemic trajectory of BMW.

Hospital-level policy changes: Within healthcare facilities, infection control policies were updated to explicitly include waste management as a critical component of COVID-19 safety. Many hospitals established new *SOPs* for waste handling in COVID units, which were not part of their pre-2020 protocols. For example, policies now specify that **all staff in COVID areas must be trained in waste segregation** and wear appropriate PPE while handling waste, whereas before this might have been limited to housekeeping staff. Hospitals also introduced *accountability measures*: Some formed internal waste management committees to do frequent audits during the pandemic, a practice that might continue. **Documentation requirements** became stricter; instead of a generic log of waste, some hospitals required noting waste output by each ward daily, enabling closer monitoring.

The pandemic has fundamentally transformed the internal culture of healthcare facilities. Perhaps the most significant change is the solidification of the mindset that views waste as a core pillar of IPC. Before 2020, waste management was often regarded as merely a logistical task relegated to the background. Today, hospital leadership acknowledges it as a crucial frontline defense for the safety of both patients and staff. As a result, hospitals are making more permanent structural changes: Increasing budgets for waste departments, hiring specialized waste managers, and strategically shifting toward on-site treatment. By investing in small-scale units like autoclaves, hospitals are protecting themselves from the external service disruptions that were prevalent during the early days of lockdowns.

At the national and state levels, the response was swift and largely driven by emergency mandates. Many countries leveraged disaster management laws to bypass bureaucratic delays, effectively "upgrading" their waste management rules overnight. A prime example is India, where the Environment Ministry circulated CPCB guidelines that became enforceable directives under the **Environment Protection Act** (Bhar et al., 2022). What began as temporary measures, such as mandatory double bagging and the use of specialized tracking apps, are now being cemented as permanent "best practices." In several Indian states, pollution control boards are already advocating for these COVID era protocols to remain the standard indefinitely, recognizing that the rigor required for a pandemic actually improves general waste safety across the board.

Another policy change was in **reporting and transparency**. Some countries, like India, now require daily or weekly COVID biomedical waste figures to be reported publicly (Jindal & Sar, 2023). This level of public reporting of biomedical waste statistics was unprecedented prepandemic. It has increased transparency and could be extended to general biomedical waste reporting beyond the pandemic, improving oversight.

In nations such as the United Kingdom and the United States, regulators initially managed the waste surge through pragmatic, temporary "relaxations." These included extending on-site storage durations or permitting alternative disposal methods, always under the watchful eye of environmental agencies. While these emergency measures have been revoked as the pandemic wanes, the experience served as a vital "stress test." It prompted a total overhaul of contingency plans; the US EPA has already refreshed its infectious waste guidelines, and the EU is currently developing robust contingency frameworks for future health emergencies based on these COVID-19 insights. One of the most significant policy victories was the formal recognition of waste management as an **essential service**. This shift ensures that in future crises, waste continuity is built into the blueprint of emergency planning, ensuring that lockdown exemptions, fuel priorities, and transport passes for waste crews are guaranteed rather than negotiated mid-crisis. The pandemic also forced a long-overdue spotlight on the informal waste sector, particularly in developing nations. Recognizing that informal waste pickers are the "invisible" backbone of the system, several governments integrated them into vaccination drives and relief programs to ensure system safety. This move has opened a window for the permanent formalization of this workforce, potentially transforming a marginalized group into a protected, recognized part of the public health infrastructure.

A New International Mandate ("Climate-Smart" Waste): On the global stage, we are seeing a fundamental shift in priority. The WHO's 2022 call for reform was a watershed moment, reframing waste management not just as a logistical burden, but as a pillar of health system strengthening (Maalouf et al., 2022; Mohamed et al., 2023). The rise of "climate-smart healthcare" now explicitly champions waste management,

Table 5.5 Key policy and practice changes in BMW management: pre-COVID versus post-COVID.

Aspect	Pre-COVID norms	During/post-COVID changes
Waste segregation	Emphasized in guidelines but compliance variable; some mixing of infectious and general waste occurred Goswami et al. (2021)	Stricter enforcement in practice. COVID isolation areas implemented *100% segregation* with dedicated bins, Ganguly and Chakraborty (2021). Noncompliance is less tolerated due to high risks. Segregation of infectious waste became nearly universal in COVID-designated facilities
Handling and storage	Routine daily collection; short-term storage on-site (often improvised in some facilities)	*Frequency increased:* Many hospitals moved to 2 3 collections per day in COVID wards. *Dedicated storage:* Separate, labeled storage areas for COVID waste became policy, Behera (2021). Emphasis on disinfection of storage surfaces and vehicles after each run
Waste worker safety	Often neglected; minimal PPE use by waste handlers; training is limited	Elevated priority. Policies mandated **PPE for all waste handlers** (gloves, masks, etc.) and training on donning/doffing Bagwan (2023). Many countries classified waste workers as *frontline personnel*, giving them early access to vaccines and resources. Likely to continue postpandemic with better occupational health policies

(continued on next page)

Regulatory oversight	Periodic reporting (monthly/annual) by hospitals; inspections are infrequent in some regions	*Real-time reporting:* Daily waste data reporting to authorities in some countries Jindal and Sar (2023). Regulators are more vigilant, with surprise inspections focused on COVID waste. Post-COVID, regulators aim to maintain tighter oversight (e.g., through electronic manifests)
Infrastructure and tech	Reliance on incineration (central or on-site) and some autoclaving; incremental upgrades	*Surge capacity and new tech:* Rapid deployment of additional incinerators, and introduction of mobile or small-scale units (microwave, autoclave). Accelerated approval of **new technologies** (plasma, etc.) under emergency use. Postpandemic, these technologies are becoming part of the standard toolkit
Waste volume reduction	Modest efforts at waste minimization (e.g., reusable linens in some hospitals), but the single-use culture is growing	Recognized as crucial after facing a waste surge. **Policies now encourage reusable options and rational use of PPE** World Health Organization (2022). Many health systems are re-evaluating which disposables can be safely reduced to cut waste volume long-term
Global guidance	Available but in the background; not a top priority in global health discourse	Front and center. WHO and UNEP issued urgent guidance Singh et al. (2022), making waste management a key part of pandemic response. Going forward, expect stronger global standards and support for countries to meet them (treated as part of health emergency preparedness)

prioritizing nonburning technologies and recycling to reduce the environmental footprint of medical care.

Cooperation and harmonization: We have also seen a new precedent for intercountry cooperation. Japan's donation of autoclaves to Southeast Asian nations during the pandemic suggests that waste infrastructure is now rightfully viewed as a critical component of global health aid packages. Furthermore, because pathogens do not respect borders, there is a renewed push for global standard harmonization. The Basel Convention and the WHO are currently integrating pandemic-specific lessons into their technical guidance (the "Blue Book"), agreeing that poor waste management in one region creates a ripple effect of environmental and health risks for others. As synthesized in Table 5.5, these changes reflect a profound evolution from the "pre-COVID" status quo to a more resilient, proactive future.

The central theme emerging from this crisis is that COVID-19 acted as a forced accelerator for long-overdue improvements. While these operational and policy shifts were born out of absolute necessity, they align perfectly with the "gold standard" experts have advocated for decades: More precise segregation, enhanced worker protections, data-driven oversight, and the adoption of modern, sustainable treatment technologies. Our primary challenge now is to ensure these gains are not lost as the immediate sense of urgency fades. We must "cement" these improvements into the permanent fabric of healthcare administration to prevent a regression to prepandemic complacency.

We are already seeing promising signs of this transition. In India, for example, the success of COVID-specific waste tracking apps has sparked a broader regulatory conversation about mandating barcode-based tracking for *all* categories of biomedical waste (Ojha et al., 2022). Such a move would be a massive leap forward for institutional accountability. Similarly, international bodies are now weaving pandemic-specific waste protocols into standard healthcare accreditation, ensuring that "preparedness" is no longer an elective but a requirement. The policy pivots of the last few years are laying a more durable foundation for the global biomedical waste landscape. We are witnessing a fundamental paradigm shift: Moving from a reactive, crisis-mode stance to a proactive, resilient strategy. By embedding these modifications into our everyday operations, we do more than just manage the refuse of today; we fortify the healthcare systems of tomorrow against whatever public health emergencies may follow.

5.7 Conclusion

The COVID-19 pandemic did far more than place pressure on hospitals and healthcare workers; it quietly but fundamentally reshaped how biomedical waste is understood and managed across the world. What had long remained on the margins of health planning suddenly became impossible to ignore, visible in hospital corridors filled with color-coded waste bags and in national systems struggling to cope with volumes never previously anticipated. Weaknesses emerged at every level, from day-to-day hospital practices to regional treatment capacity and policy oversight, yet the crisis also served as a moment of reckoning. As this chapter has shown through global experiences and South Asian case studies, the unprecedented surge in infectious waste forced a long-overdue reassessment of BMW systems. At the global level, the pandemic marked a decisive turning point, as biomedical waste could no longer be dismissed as a secondary technical concern and instead emerged as a core pillar of public health, occupational safety, and environmental protection. When waste volumes multiplied rapidly, international bodies such as the **WHO** did not simply advocate for expanded collection or disposal capacity, but encouraged a broader shift toward sustainability, including reusable PPE and cleaner, nonburning sterilization technologies that reduce long-term environmental harm (Maalouf et al., 2022; Mohamed et al., 2023). In South Asia, the experience highlighted both preparedness and vulnerability. In India, regulatory frameworks that were often questioned during normal times demonstrated their value by enabling swift policy adaptations, reinforcing the importance of investing in waste systems before emergencies occur (Bhar et al., 2022; Khosla et al., 2022). Elsewhere, including in Bangladesh, the crisis exposed significant gaps, but the difficult lessons learned are now driving improvements in infrastructure, workforce capacity, and regulatory enforcement (Behera, 2021). Within hospitals, the pandemic dissolved any artificial boundary between infection control and waste disposal, as healthcare workers learned that improper waste handling is not merely a logistical lapse but a direct clinical risk. This realization is likely to leave a lasting imprint on hospital culture, with dedicated infection control teams, routine staff training, and clearer accountability becoming standard expectations rather than optional measures. Technologically, BMW is entering a more mature phase, where digital tracking, IoT-based monitoring, and advanced sterilization are no longer aspirational ideas but practical tools needed to address chronic challenges such as waste overflow and pollution-intensive incineration practices (Capoor & Parida, 2021; Sharma et al., 2020). Ultimately, the post-COVID period offers a genuine opportunity to build more resilient systems, provided the urgency of the crisis is translated into sustained commitment. For researchers, policymakers, and healthcare providers alike, the task ahead is to transform these hard-won lessons into everyday standards of care that protect the environment, safeguard public health, and uphold the dignity of the workers who sustain healthcare systems, not only in recovery from the last crisis but in preparation for those yet to come.

Ultimately, the post-COVID period represents a genuine opportunity to build back better. The pandemic exposed vulnerabilities at an enormous cost, but it also demonstrated that rapid and focused improvement is possible when urgency aligns with political and institutional commitment. If this momentum can be sustained, COVID-19 may be remembered as the moment when global biomedical waste safety began to change for the better. For researchers, the work is only beginning. There is a need to understand the long-term ecological impact of billions of discarded masks and to study how human behavior around waste segregation changes in high-stress environments. For policymakers and healthcare providers, the task ahead is to convert these hard-won lessons into everyday standards of care that respect the environment and uphold the dignity of the workers who keep healthcare systems functioning. By doing so, the goal is not only to recover from the last crisis but to prepare wisely and responsibly for the next one.

References

Abbasi, S., & Sıcakyüz, Ç. (2024). A review of the COVID-19 pandemic's effects and challenges on worldwide waste management.

Adelodun, B., et al. (2021). Insights into hazardous solid waste generation during COVID-19 pandemic and sustainable approaches.

Akcam, F.Z., et al. (2023). How medical waste management was affected by the COVID-19 pandemic in hospitals.

Al-Emam, R., & Al-Yousfi, B. (2021). Waste management of used PPE during the COVID-19 pandemic in the Eastern Mediterranean Region.

Ali, M., et al. (2022). An environmentally friendly solution for waste facial masks recycled in construction materials.

Ali, S.A., & Parvin, F. (2022). Examining challenges and multi-strategic approaches in waste management.

Aliyi, A.A., et al. (2024). Prevalence of injury and utilization of PPE among building construction workers.

Almulhim, A.I., et al. (2021). Consequences of COVID-19 pandemic on solid waste management.

Al-Omran, K., et al. (2021). Estimation of COVID-19 generated medical waste in the Kingdom of Bahrain.

Amin, F., et al. (2024). Utilization of discarded face masks in combination with recycled concrete aggregate.

Bagwan, W.A. (2023). An investigation of the bio-medical waste produced in India during the COVID-19 pandemic.

Baniak, L.M., et al. (2025). Priorities for Healthcare Systems Emergency Preparedness in Inpatient Settings Informed by COVID-19.

Bank A.D. (2020). Managing infectious medical waste during the COVID-19 pandemic.

Barua, U., & Hossain, D. (2021). A review of the medical waste management system at Covid-19 situation in Bangladesh.

Behera, B.C. (2021). Challenges in handling COVID-19 waste and its management mechanism: A Review.

Bhar, A., et al. (2022). The influence of COVID-19 pandemic on biomedical waste management, the impact beyond infection.

Bizuneh, Y.B., et al. (2024). Assessment of knowledge, attitude, and practice regarding medical waste management.

Budati, S., et al. (2024). Management of COVID-19 medical waste based on pyrolysis-swot analysis.

Capoor, M.R., & Parida, A. (2021). Current perspectives of biomedical waste management in context of COVID-19.

Celitron Medical Technologies. (2019). Biomedical waste management colour coding for beginners.

Central Pollution Control Board (India). (2020). Guidelines for Handling, Treatment and Disposal of Waste Generated during COVID-19.

Česonienė, L., et al. (2022). Analysis of Hazardous Plastic Waste Generated During the COVID-19 Pandemic in Lithuania.

Chand, S., et al. (2021). Updates on biomedical waste management during COVID-19: the Indian scenario.

Das, A.K., et al. (2021). COVID-19 pandemic and healthcare solid waste management strategy–A mini-review.

Dehal, A., et al. (2022). Biomedical waste generation and management during COVID-19 pandemic in India.

Faisal, G.M., et al. (2021). Challenges in medical waste management amid COVID-19 pandemic in a megacity Dhaka.

Ganguly, R.K., & Chakraborty, S.K. (2021). Integrated approach in municipal solid waste management in COVID-19 pandemic.

Gill, Y.Q., et al. (2022). Review of hospital plastic waste management strategies for Pakistan.

Goswami, M., et al. (2021). Challenges and actions to the environmental management of Bio-Medical Waste in India.

Gupta, P.P., et al. (2023). The Efficient Disposal of Biomedical Waste Is Critical to Public Health.

Hackett, D.P. (1989). Assessment of the Basel Convention on the control of transboundary movements of hazardous wastes.

Hamid, A., et al. (2013). Striving for scientific management of medical waste: challenge for Dhaka City.

Hantoko, D., et al. (2021). Challenges and practices on waste management and disposal during COVID-19 pandemic.

Haque, M.S., et al. (2021). Coronavirus disease 2019 (COVID-19) induced waste scenario: A short overview.

Hossain, I., et al. (2020). Pandemic COVID-19 and biomedical waste handling: a review study.

Hussein, L., et al. (2022). A review on waste management and compost production in the MENA region.

Idrees, M., et al. (2022). Recycling of waste facial masks as a construction material.

Ilyas, S., et al. (2020). Disinfection technology and strategies for COVID-19 hospital and bio-medical waste management.

Ishaq, A., et al. (2025). Smart waste bin monitoring using IoT for sustainable biomedical waste management.

Janik-Karpinska, E., et al. (2023). Healthcare waste a serious problem for global health.

Jebaranjitham, J.N., et al. (2022). Current scenario of solid waste management techniques and challenges in Covid-19.

Jindal, M.K., & Sar, S.K. (2023). Medical waste management during COVID-19 situation in India: Perspective towards safe environment.

Kankanamge, C.E., et al. (2020). Personal protective equipment use during COVID-19 and associated waste management in households.

Kanwar, V.S., et al. (2023). An overview for biomedical waste management during pandemic like COVID-19.

Kashyap, S., et al. (2020). Waste quarantine to reduce COVID-19 infection spread.

Khosla, R., et al. (2022). Upsurge in biomedical waste due to COVID-19 in India: A statistical correlation.

Kim, M.Y., et al. (2024). Knowledge, Attitude, and Practice of Medical Waste Management in Medium-sized Hospital.
Kothari, R., et al. (2021). COVID-19 and waste management in Indian scenario: challenges and possible solutions.
Kudli, L.P., et al. (2021). Biomedical waste management during the COVID-19 Pandemic–Indian Scenario.
Kulkarni, B.N., & Anantharama, V. (2020). Repercussions of COVID-19 pandemic on municipal solid waste management.
Kuppusamy, P.P., et al. (2022). Accumulation of biomedical waste during the COVID-19 pandemic.
Liang, Y., et al. (2021). Repercussions of COVID-19 pandemic on solid waste generation and management strategies.
Liu, H.C., et al. (2015). Evaluating health-care waste treatment technologies using a hybrid multi-criteria decision making model.
Luhar, I., et al. (2022). Challenges and impacts of COVID-19 pandemic on global waste management systems: a review.
Lyu, Y.Y., et al. (2025). Medical waste management practices in primary, secondary, and tertiary hospitals in Anhui, China.
Maalouf, A., et al. (2022). Medical Waste Management in Lebanon and Impact of COVID-19.
Manekar, S.S., et al. (2022). Challenges and measures during management of mounting biomedical waste in COVID-19 pandemic: an Indian approach.
Manupati, V.K., et al. (2021). Selection of the best healthcare waste disposal techniques during and post COVID-19.
Mejjad, N., et al. (2021). Disposal behavior of used masks during the COVID-19 pandemic in the Moroccan community.
Mijakoski, D., et al. (2020). Personal protective equipment in health workers during coronavirus Disease-19 outbreak.
Mohamed, N.H., et al. (2023). Modernizing medical waste management: unleashing the power of IoT.
Mohamed, N.H., et al. (2024). Waste 4.0: transforming medical waste management through digitalization.
Mohammad, A., et al. (2020). Discussion on 'Challenges, opportunities, and innovations for effective solid waste management'.
Mondal, R., et al. (2022). COVID 19 Pandemic and biomedical waste management practices in healthcare system.
Najafighodousi, A., et al. (2024). Recycling facemasks into civil construction material.
Nayan, R., et al. (2021). Vaccination Drive, Vaccine Diplomacy for COVID-19 Pandemic: Indian Perspective.
Ojha, P.C., et al. (2022). Overcoming challenges due to enhanced biomedical waste generation.
Ojha, P.C., et al. (2022). Overcoming challenges due to enhanced biomedical waste generation during COVID-19.
Patil, A. D., & Shekdar, A. V. (2001). Health-care waste management in India. *Journal of environmental management*.
Platon, V., et al. (2020). Challenges in adequate management of hazardous medical waste in Romania.
Rajak, R., et al. (2022). Assessment of bio-medical waste before and during the emergency of novel Coronavirus disease pandemic in India.

Ramteke, S., & Sahu, B.L. (2020). Novel coronavirus disease 2019 (COVID-19) pandemic: considerations for the biomedical waste sector.
Ranjan, M.R., et al. (2020). Medical waste generation during COVID-19 pandemic and its management: an Indian perspective.
Rankin, N.M., et al. (2024). COVID-19 research response to immediate demands: setting priorities with key stakeholders.
Sawant, D., & Kashte, S. (2024). COVID-19 Biomedical Waste Disposal Management Challenges, Solutions and Prospects.
Selvan Christyraj, J.R., et al. (2021). Impact of biomedical waste management system on infection control.
Shammi, M., et al. (2021). The escalating biomedical waste management to control the environmental transmission of COVID-19.
Sharma, H.B., et al. (2020). Challenges, opportunities, and innovations for effective solid waste management.
Shbaklo, N., et al. (2021). Infection control in the era of COVID-19: a narrative review.
Shekoohiyan, S., et al. (2021). People knowledge about waste management during the outbreak of COVID-19 in Tehran.
Siddiqui, A., et al. (2021). An overview of procurement, pricing, and uptake of COVID-19 vaccines in Pakistan.
Singh, D., et al. (2022). Mask consumption and biomedical waste generation rate during Covid-19 pandemic.
Singh, E., et al. (2022). Solid waste management during COVID-19 pandemic: Recovery techniques and responses.
Singh, H., et al. (2024). Harnessing the foundation of biomedical waste management for fostering public health.
Singh, N., et al. (2020). COVID-19 waste management: effective and successful measures in Wuhan, China.
Soares, S.R., et al. (2013). Applications of life cycle assessment and cost analysis in health care waste management.
Srivastava, A., et al. (2024). Can pyrolysis handle biomedical wastes?.
Sujon, H., et al. (2022). Medical waste management: an assessment of district-level public health facilities in Bangladesh.
Tripathi, A., et al. (2020). Challenges, opportunities and progress in solid waste management during COVID-19 pandemic.
Ugandar, R.E., et al. (2023). Hospital waste management using internet of things and deep learning.
Um, N., et al. (2023). An improved strategy for effectively managing the transboundary movement of waste based on the Basel Convention.
UNEP. (2020). Waste management an essential public service in the fight to beat COVID-19.
Voudrias, E.A. (2024). Management of COVID-19 healthcare waste based on the circular economy hierarchy.
Wang, J., et al. (2024). Survey on medical waste disposal in medical and healthcare institutions in Hefei City.
World Health Organization. (2020). Water, sanitation, hygiene, and waste management for the COVID-19 virus: interim guidance
World Health Organization. (2022). Tonnes of COVID-19 health care waste expose urgent need to improve waste management systems.

World Health Organization. (2024). Health-care waste Fact Sheet.

Yang, L., et al. (2021). Emergency response to the explosive growth of health care wastes during COVID-19 pandemic in Wuhan.

Zarocostas, J. (2022). WHO concerned over COVID-19 health-care waste. *Lancet*.

CHAPTER

Recent advances in healthcare waste treatment: Innovations, challenges, and environmental sustainability

6

Mukta Mayee Kumbhar, Suman Jagatee and Sujata Priyadarshini
Department of Life Sciences, Rama Devi Women's University, Bhubaneswar, Odisha, India

6.1 Introduction

Healthcare facilities produce biomedical waste (BMW) while providing healthcare services, and therefore, managing healthcare waste remains a serious challenge in most developing countries. These BMWs are generated by facilities like blood banks, biological research, and other medical facilities during the process of diagnosing, treating, and vaccinating humans or animals. Improper handling can have catastrophic consequences for human health and the environment. The BMW increases dramatically during infectious disease outbreaks; as a result, management must take extra precautions to prevent unsettling effects. The rapid rise in the production of BMW has been attributed to the severe acute respiratory syndrome, also known as coronavirus illness (COVID-19), which has adversely affected the environment and public health. It is also indicated that the total COVID-19 generation of BMW is approximately 101 MT every day. In December 2020, the top five states producing BMW in India were Maharashtra, Kerala, Gujarat, Andhra Pradesh, and Delhi. In addition to polluting the environment, improper disposal of medical waste can lead to the spread of infectious diseases like cholera, typhoid, HIV/AIDS, hepatitis, and respiratory complications. These diseases are primarily brought on by scavenging medical waste or reusing discarded medical equipment, according to reports from various nations. BMW production increased sixfold during the COVID-19 pandemic compared to the situation before the outbreak. Coronavirus or COVID-19 has been documented to infect about 657 million people worldwide and to have resulted in 6,681,433 fatalities (World Health Organization, 2022). According to reports, every day BMW generates in India is about 1.5–2 kg per bed (Singh et al., 2022). Between June 2020 and August 2022, when the COVID-19 pandemic struck India, Kerala and Maharashtra emerged as the country's top two producers of BMW, producing 74,084

Biomedical Waste Management in the COVID-19 Era: A Global Scenario.
DOI: https://doi.org/10.1016/B978-0-443-21841-5.00009-6

tons and 39,992 tons respectively. However, there was no appreciable correlation found between the quantity of BMW generated and the number of cases of the disease, which might be due to proper handling of BMW in those states (Kumbhar et al., 2024a). A significant rise in the production of BMW (2.5–4.0 kg per bed) was reported post-COVID when compared to the pre-COVID situation (Sharma et al., 2020).

According to information available, India produced 220,595.85 tons of BMW between July 2020 and June 2022 (Central Pollution Control Board, 2024). However, the country generated 705 tons of BMW waste per day in 2022, according to the annual report information. This is fewer than the 764 tons of BMW generated per day in 2021. In 2022, 59 tons (or around 8%) less BMW was produced daily than in 2021; this difference could be attributed to a rise in BMW production during the COVID-19 epidemic. There has been a 5% increase in the overall number of healthcare institutions in 2022 compared to 2021.

The BMW Management and Handling Rules were formulated on July 20, 1998. Further, in accordance with the Environment (Protection) Act, 1986, the Ministry of Environment and Forests (MoEF) replaced the 1988 BMW Rules with the new ones on March 28, 2016, to enhance the segregation, quantification, disposal method, transportation, and treatment of BMW, as it can pollute the air, water, and land if improperly managed. Effective BMW disposal requires a lot of cooperation as well as steadfast government backing for infrastructure and financial growth. Devoted healthcare facilities and personnel are also important.

To some extent, the implementation of new waste segregation and management rules was successful. However, there is a lack of ground-level management for solid and medical wastes (Bansod & Deshmukh, 2023). Government agencies developed some monitoring and evaluation techniques to keep tabs on garbage production and use appropriate disposal strategies in collaboration with local people and municipalities. Furthermore, it is imperative that BMW be properly and consistently monitored. This study provides a comprehensive assessment of appropriate plans and procedures for BMW treatment methods to build an eco-friendly, clean, and green environment.

6.2 Types of biomedical waste

Hospital waste typically comes in two varieties: Hazardous waste and nonhazardous waste. Annual composition of BMW includes 85% nonhazardous waste, and the remaining 15% is hazardous waste (Fig. 6.1). To neutralize the associated biohazards, infectious waste needs to be treated specifically since it contains microbes that could cause an infectious disease in a vulnerable host (Usman & Elizabeth, 2020).

6.3 Hazardous biomedical waste

Hazardous BMW includes sharps, human body parts, blood, body fluids, feces, chemical waste, pharmaceutical waste, and medical devices, microbes (fungi, viruses, and bacteria), etc. The contagious pathogens are transported by surgical organs or

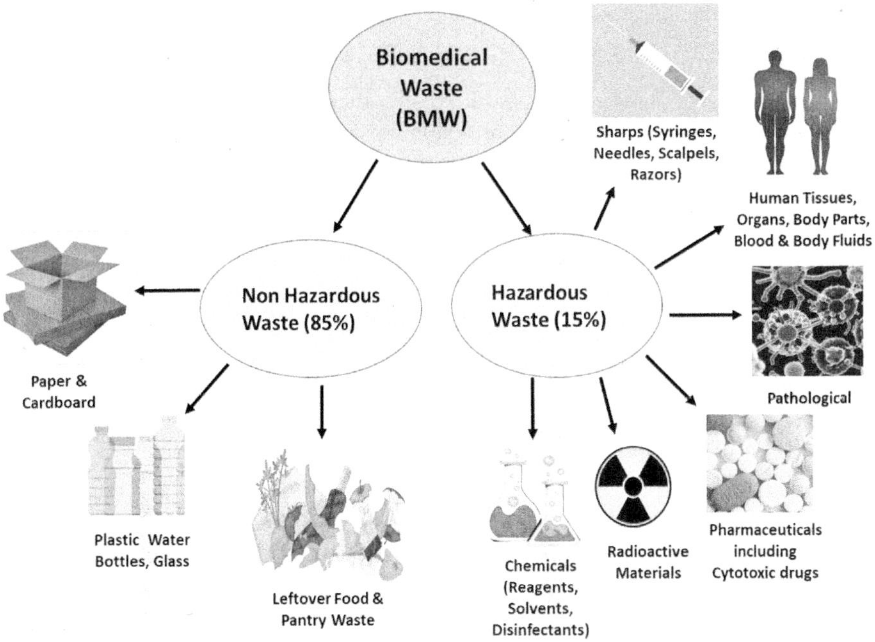

FIGURE 6.1

Schematic representation of various types of biomedical waste (BMW).

biopsy material, including blood and various body fluids, infecting other susceptible hosts. In addition to this, some of the waste products are severely contaminated with harmful chemicals, making them biohazardous. Moreover, pharmacological waste, which includes unused and expired medications, poses a threat to the environment due to the presence of various harmful and biologically active chemical derivatives. Their metabolic processes are vulnerable to a range of erratic biochemical reactions in the exposed atmosphere, resulting in the production of advanced hazardous compounds that represent a threat to the environment (Dehal et al., 2022).

People around the world were advised to wear face masks due to the COVID-19 pandemic. The World Health Organization estimated that 89 million disposable face masks were used daily throughout the world (WHO 2020a, 2020b). According to Singh et al. (2020), the protective N95 mask and protective gear, such as gloves and face shields, contain potentially harmful substances like polypropylene and Tyvek that can linger in the environment for a long time and release dioxin and other harmful substances. The microplastics generated by face masks have the potential to become a catastrophe if they are not properly addressed and managed. It is important to use, dispose of, recycle, and handle these facemasks carefully to reduce contamination in both aquatic and terrestrial environments.

6.4 Nonhazardous biomedical waste

Unused food, pantry waste, cardboard, plastic, thermocycling, and paper produced during various medical operations are examples of nonhazardous BMWs, as long as they are not contaminated. The majority of nonhazardous BMWs are caused by pantry waste and unconsumed food. They are not hazardous, and they do not need to be preprocessed in any particular way before being thrown out or recycled. According to Datta et al. (2018), they are typically carried via a number of municipal solid waste management systems that are overseen by the corresponding urban or rural management. They can be used to make fertilizer or biogas. Isolation at the source is only monitored to reduce the processing and removal load because their quantity is significantly larger than that of hazardous BMW. Effective segregation can significantly lower BMW's overall management expenses (Antoniadou et al. 2021).

6.5 Techniques for the management of biomedical waste

BMW poses a risk to the health and safety of an individual. Furthermore, the various forms of BMW have increased concerns regarding waste management, including shipping, storage, and disposal. Innovative methods have been created recently to shield humans from dangerous and contagious substances derived from BMW. Apart from eliminating all infections and organic hazardous waste, it also significantly lowers the amount of trash produced (Manupati et al., 2021). In hospitals and labs, numerous methods are currently in use to treat BMW; some are listed as follows:

i. Landfill disposal
ii. Autoclave treatment
iii. Chemical disinfection
iv. Incineration
v. Pyrolysis technique
vi. Microbial bioremediation

6.5.1 Landfill disposal

This is the easiest way to get rid of the BMW after the chemical disinfection procedure. Biomedical solid waste is dumped in landfill sites, which are 2–3 m deep pits (Kumbhar et al., 2024a). Waste decomposition is accelerated by a variety of biological, physical, and chemical reactions (Kothari et al. 2021). Organic garbage that has been deposited in landfills quickly undergoes degradation and decomposition, whereas plastics and other polymer wastes can take 10–100 years to decompose (Thiounn & Smith, 2020).

6.5.2 Autoclave treatment

The most popular and effective wet thermal method for handling medical waste is autoclaving, which uses high-temperature steam under pressure and vacuum. Due to its simplicity, affordability, and decreased air pollution, this approach is frequently used for the sterilization of BMW. Wastes are sterilized or disinfected using this approach before being dumped in a landfill. Waste bags are put in a chamber, and steam is added for a predetermined amount of time at a predetermined temperature and pressure, which ensures that pathogenic microbes will be destroyed. It is appropriate for handling glassware, sharp wastes, medical equipment that can be reused, solid healthcare wastes, and used personal protective equipment. However, autoclaves are not appropriate for cytotoxic, pathological, or other hazardous chemical wastes (Mishra et al., 2023).

6.5.3 Chemical disinfection

In healthcare institutions, general disinfectants such as ozone, sodium hypochlorite, liquid chlorine, chlorine dioxide, and UV radiation are employed to sterilize surfaces. These disinfectants are safe, odorless, tasteless, colorless, noncorrosive, and readily soluble in water, which do not leave any residual hazards after disinfection, and are not impacted by any physical or chemical factors (Chen & Yang, 2016; Tudu et al., 2026). Infectious germs are either killed or rendered inactive during the disinfection process. Before receiving chemical treatment, the BMWs should be appropriately shredded. Typically, chemical disinfection is only used for liquid and chemical waste. However, the effectiveness of the method depends on the kinds of contamination present in the infected BMW and allows BMW to undergo oxidation, reduction, precipitation, and pH neutralization, turning it into less dangerous substances (Lindmark et al., 2022). Even if it effectively destroys germs, waste volume might not be greatly reduced. Aerosols are released when trash is chemically treated, and these aerosols can enter the alveoli and increase the chance of developing lung cancer in humans (Ilyas et al., 2020). However, this technique may be taken into consideration when hospital waste is minimal.

6.5.4 Incineration

This is an easy, safe, efficient technique, and installing incinerators is required in all healthcare facilities. The incinerator's exit temperature is more than 800°C. In addition to killing microorganisms entirely, this high-temperature has the potential to burn most organic materials and turn them into inorganic dust. The amount of solid trash might be cut by about 85%–90% after incineration (Ghodrat et al., 2017). Since this aerated process generates several pollutants, including dioxins, furans, and others that contribute to the greenhouse effect and global warming, further treatment is necessary (Ilyas et al., 2020). The installation of incinerators and secondary treatment plants is costly, but due to their effectiveness, they are regarded as the best option

(Ojha et al. 2022). BMW generates a huge amount of plastic waste along with other solid and liquid waste. Polymers included in plastics are typically resistant to explosives and oxygen heating. Many developing countries, including Bangladesh, India, Pakistan, Brazil, Argentina, and Peru, commonly handle their BMWs by burning them (Manzoor & Sharma, 2019). Before being burned, the waste should, nevertheless, get a rudimentary disinfecting treatment.

6.5.5 Pyrolysis technique

This method is frequently used for dry and solid waste materials that are not recyclable. It is based on high-temperature combustion (900–1200°C); this method eliminates pathogens (Datta et al., 2018). It is a good way to reduce the majority of waste, but there are drawbacks as well. The main ones are the smoke, odor, and ongoing setup, as well as the fact that the high heat release makes it difficult to recover carbon fibers. Furthermore, incineration does not damage polymers or metals. Pyrolysis is a rapid thermochemical process that can be utilized as an alternative method to treat COVID-19 waste since it breaks down municipal solid waste at high temperatures without oxygen (Kumbhar et al., 2024b). A variety of pyrolysis methods are employed, including laser-based pyrolysis, plasma pyrolysis, pyrolysis oxidation, and induction-based pyrolysis. The best feature is that it releases clean exhaust steam, known as syn-gas, after processing and gets rid of the toxic gaseous vapors that build up inside the chamber. This makes it the most advanced and safest disposal method (Ojha et al., 2022).

6.5.6 Microbial bioremediation

The treatment of BMW is a worldwide concern, and research is being conducted to identify low-cost and eco-friendly strategies for handling this kind of waste because the current nonbioremediation approaches are seen to be expensive and unfriendly to the environment. The inadequate handling of radioactive waste, plastics, and heavy metals in particular is a challenge that drives the need for more environmentally friendly options for their ultimate disposal (Kumbhar et al., 2026). Bioremediation has the ability to successfully and economically recover polluted areas. It takes a multidisciplinary approach to understand the biology and mechanism of bioremediation. The target pollutant is broken down by biostimulation, bioaugmentation, bioleaching, biosorption, bioaccumulation and natural attenuation by microorganisms. But the lack of understanding of the variables controlling the development and metabolism of microorganisms in contaminated environments frequently limits their application. The bioremediation process converts hazardous wastes into less toxic or nontoxic compounds by using naturally occurring microorganisms. Because they can be easily handled and proliferate more quickly than other agents of bioremediation, microbes are used more than others. Various taxa of bacteria, fungi, and algae have been used to clean up a variety of environmental contaminants (Ayilara & Babalola, 2023; Jagatee et al., 2026). Hospital-generated BMW can be biologically handled since it is a mixture of liquid and solid materials with a high concentration of organic contaminants.

This method is thought to be an alternative to the chemical and physical methods of degradation of BMW, which are more affordable and environmentally friendly. By altering the environmental circumstances that promote the active growth of microbes toward the degradation of specific contaminants, particularly BMWs, it is used to remediate polluted soil and water (Dyson et al., 2018). Because of their innate physiological, metabolic, and genetic characteristics, microorganisms are the best option for remediating contaminants found in soil and groundwater. These days, a range of sophisticated molecular methods has been applied to investigate the microbial community in the region that requires bioremediation (Priyadarshini et al., 2026). Furthermore, it is advised to employ plants in phytoremediation to break down hazardous biological waste materials that are found in soils and groundwater. Both microbial remediation and phytoremediation techniques show a greater possibility for cleaning up different kinds of pollutants found in hospital waste. This method is very cost-effective, environmentally friendly, and capable of managing even radioactive waste from hospitals (Lopez-Fernandez et al., 2021). This method is very cost-effective, environmentally friendly, and capable of managing even radioactive waste from hospitals.

6.6 Adverse impact of biomedical waste on the environment and living organisms

An enormous amount of medical waste has been released into the environment since the pandemic started. These medical wastes adversely impact the environment and organisms in multiple ways, which are schematically represented in Fig. 6.2. Chlorine compounds are used in disinfectants, and excessive use of chlorine-based sprays eventually accumulates in water bodies. When chlorine-based disinfectants are added to aquatic media, they react with the organic materials that are naturally present, posing a concern. The chlorine residue produces more toxic disinfection byproducts that can potentially harm aquatic life, including trihalomethane (THM), haloacetic acid (HAA), haloacetaldehyde (HAL), haloacetonitrile (HAN), and halonitromethane (HNM) (Furst et al., 2019). An estimated 7000 tons of medical waste are produced daily as a result of the COVID-19 outbreak, with abandoned masks accounting for a sizable amount of this waste. Healthcare waste like personal protective equipment (PPE) and facemasks breaks down into microplastics (<5 mm diameter) that are transported from one place to another by rivers, seas, and contaminate both terrestrial and aquatic environments (Subramanian et al., 2023). Microplastic (MP) contamination raises concerns over the security of the global food supply as most of the aquatic organisms are vital components of the food web. Every day, more than 170,000 microfibres can be released into the environment from a single, discarded face mask. Typically, synthetic polymers that are not biodegradable, including polyester and polypropylene, are used to make masks.

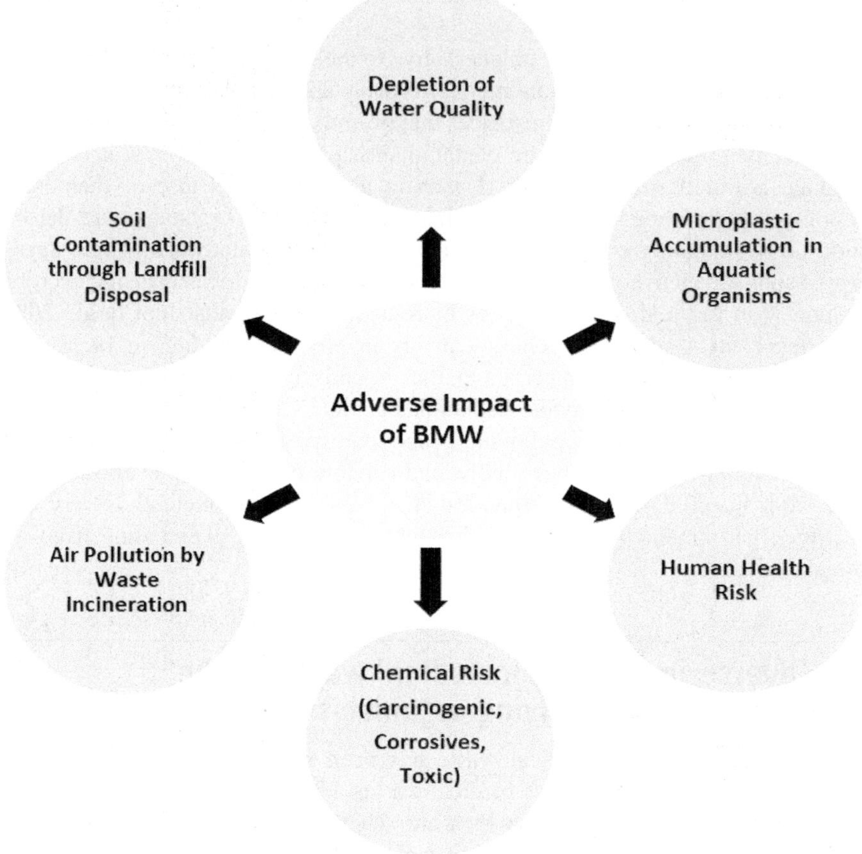

FIGURE 6.2

Illustration of the adverse impact of biomedical waste (BMW) on the environment and human health.

Without adequate management, the deposition of single-use plastic materials into aquatic environments has resulted in the accumulation of approximately 130 billion face masks and 64.8 billion gloves, which has increased the number of emerging pollutants (Silva et al., 2020). Microplastics that accumulate in fish bodies eventually finds its way into human bodies, where they can lead to diseases like cancer, lung fibrosis, reproductive problems, liver and kidney damage, respiratory tract inflammation, and hormone imbalance (Jimoh et al., 2023). Additionally, MPs can cause a variety of illnesses, including interaxial tissue (pneumothorax), inflammatory and fibrotic alterations in bronchial and peribranchial tissue (chronic bronchitis), diffuse interstitial fibrosis and fiber-inclusive granulomas (extrinsic allergic alveolitis and

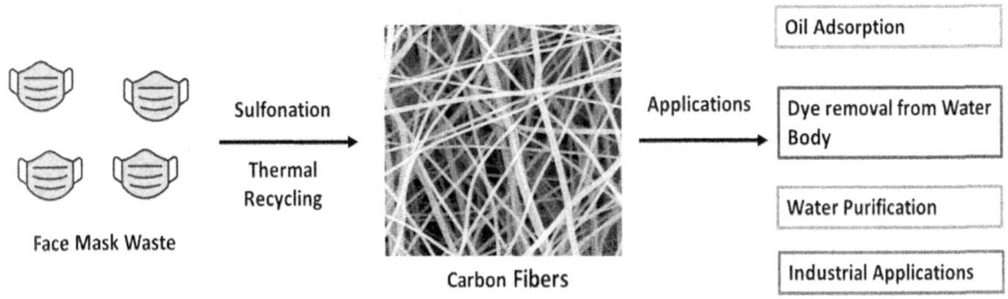

FIGURE 6.3

Recovery of carbon fibers from face mask and their applications.

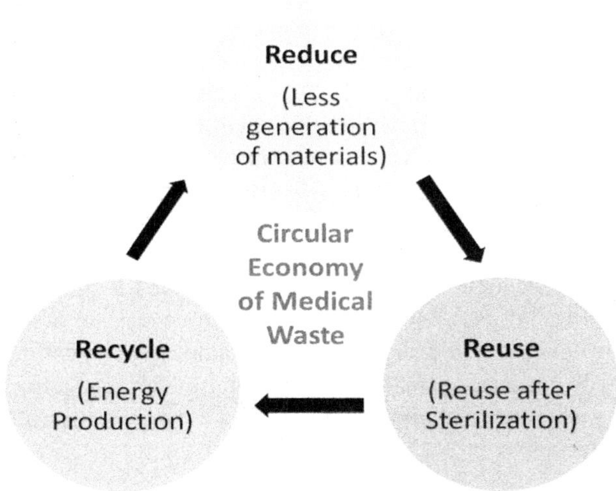

FIGURE 6.4

Approach of the 3R strategy of circular economy for the sustainable management of medical waste.

chronic pneumonia), and immediate (asthma-like) bronchial reactions. Besides, filter-feeding bivalves such as clams, mussels, and oysters are frequently exposed to plastic and typically consume MP. Microplastics smaller than 10 μm were ingested by blue mussels (Mytilus edulis), formed in the gut, and then entered their circulatory system (Sussarellu et al., 2016; Kaur et al., 2026). Besides, MP accumulation in tissue can have several harmful effects, including oxidative stress, immune deficiencies, physical harm, immunological deficiencies, genetic, neurotoxic, and metabolic malfunctions.

6.7 Recycling and recovery of carbon fibers from healthcare waste

More than 3 billion masks were discarded daily in 2020, and wearing a face mask is now recommended for those with respiratory conditions. This high rate is expected to increase in view of the new strain of COVID-19, which is probably going to continue in the years to come as a result of several respiratory illnesses. These disposable face masks are made up of polypropylene (PP), which contaminates the environment and ultimately affects the health of organisms living there. The global problem of face mask recycling needs to be addressed with practical solutions to create a sustainable environment and society. The idea of recovering carbon fiber from polyolefins, such as polyethylene (PE), was introduced by Horikiri et al. (1974). Researchers have lately looked into turning PP fibers from face masks into carbonaceous materials via the sulfonation crosslinking technique. More recently, scientists have looked into turning PP fibers from face masks into carbonaceous materials via the sulfonation crosslinking technique (Lee et al., 2022). Thermal stabilization of PP is achieved by heating it in concentrated sulfuric acid (Frank et al., 2021). The stabilized PP is then subjected to high temperatures (800–1400°C) in a N_2 environment to undergo pyrolysis and produce the final carbon product, which maintain their microscopic fiber architectures, and the yield of carbon is more than 50% (Robertson et al., 2022). Due to their superior strength, low weight, and dependability over alternative materials, carbon fibers are increasingly utilized in contemporary industrial applications. Further, these face masks are the precursor sources to create value-added fluorescent carbon fibers, which have a variety of uses, including the removal of organic pollutants and dyes from the aquatic environment. Due to their hydrophobic nature, carbon materials can be used for oil adsorption (Fig. 6.3). The carbon mask fibers can adsorb up to 14 g of mineral oil per gram of carbon fiber, demonstrating a range of adsorption capabilities for various organic solvents (Robertson et al., 2022).

6.8 Circular economy for the management of biomedical waste

Due to its potential benefits for sustainability, economic gain, and lowering the exploitation of primary resources, the circular economy concept has received a lot of attention. This concept has been widely adopted in food, plastic, and manufacturing industrial waste management (Bianchi & Cordella, 2023). Mostly, the medical waste is infectious, toxic, and hazardous, and therefore, the medical industry has been less active in the discussion and implementation of the circular economy. This is because the medical industry tends to use single-use practices. In addition to conceptualizing the green growth of the natural environment, the circular economy is an impressive idea for creating a green economy. The medical industry has grown significantly during the last 10 years, which has resulted in a notable rise in medical waste overall. To achieve sustainable development, which leads to social fairness, improves environmental quality, balances economic success, and creates a secure planet for future generations, the circular framework comprises the 3R strategies: Reduce, reuse, and recycle (Fig. 6.4). The unanticipated COVID-19 pandemic has made the problem worse, with an estimated 3.4 kg cap^{-1} d^{-1} rise in COVID-19-related medical waste globally (Kenny & Priyadarshini, 2021). Recycling helps conserve natural resources, saves rapidly dwindling landfill space, lowers pollution in all habitats, and uses less energy. Nowadays, the concept of turning garbage into energy is becoming more and more popular. However, the recycling rate decreased as a result of the contagious nature of COVID-19-related garbage. To ensure environmental sustainability during the pandemic and post-COVID crises, international organizations such as the United Nations Environment Program (UNEP) have proposed guidelines and recommendations for waste recycling mechanisms in cooperation with policymakers. Circular economy strategies, such as recycling and tough rules against plastic pollution, require an urgent and coordinated commitment.

6.9 Conclusion

Healthcare facilities produce BMW, which can be contagious and dangerous. Inappropriate handling may pose health risks. Important phases in BMW management include collection, segregation, transportation, treatment, and disposal. Bin color coding, the application of technologies like autoclaving and cremation, and consideration of environmental effects are also very important. The main objective of BMW management is to reduce the volume of medical waste and ensure its proper disposal to create a green environment. The aquatic and terrestrial ecosystems are threatened by the disposal of healthcare waste, plastics, heavy metals, and organic contaminants. Consequently, a suitable BMW management plan is regarded as the most important component of hospital sanitation and maintenance. Many reports indicate that researcher are going on to assess the physiology of microbes in mineralizing pollutants by increasing their neutralization efficiency. In addition to microbial bioremediation, phytoremediation also plays a key role in the biodegradation of medical waste. The biomass produced by phytoremediation can be utilized by commercial facilities to produce energy and heat. Thus, it has the potential to develop into a new, eco-friendly technology.

References

Antoniadou, M., Varzakas, T., & Tzoutzas, I. (2021). Circular economy in conjunction with treatment methodologies in the biomedical and dental waste sectors. *Circular Economy and Sustainability, 1*(2), 563–592. doi:10.1007/s43615-020-00001-0.

Ayilara, M. S., & Babalola, O. O. (2023). Bioremediation of environmental wastes: The role of microorganisms. *Frontiers in Agronomy, 5*, 1183691.

Bansod, H. S., & Deshmukh, P. (2023). Biomedical waste management and its importance: A systematic review. *Cureus, 15*(2), e34589. doi:10.7759/cureus.34589.

Bianchi, M., & Cordella, M. (2023). Does circular economy mitigate the extraction of natural resources? Empirical evidence based on analysis of 28 European economies over the past decade. *Ecological Economics, 203*, 107607.

Central Pollution Control Board (2024) https://cpcb.nic.in/.

Chen, M., & Yang, H. (2016). Current status of medical waste disinfection and disposal technologies. *Chinese Journal of Disinfection, 33*, 71–174.

Datta, P., Mohi, G. K., & Chander, J. (2018). Biomedical waste management in India: Critical appraisal. *Journal of Laboratory Physicians, 10*, 6–14. doi:10.4103/JLP.JLP_89_17.

Dehal, A., Vaidya, A. N., & Kumar, A. R. (2022). Biomedical waste generation and management during COVID-19 pandemic in India: Challenges and possible management strategies. *Environmental Science and Pollution Research International, 29*(10), 14830–14845. doi:10.1007/s11356-021-16736-8.

Dyson, C., Kumar, S. N., Samuel, A. M., Devipriya, H. B., & Midhila, M. S. (2018). Study of bioremediation on hospital plastic waste management. *International Journal of Management Technology and Engineering, 8*(12), 678–686.

Frank, E., Muks, E., Ota, A., Herrmann, T., Hunger, M., & Buchmeiser, M. R. (2021). Structure evolution in polyethylene-derived carbon fiber using a combined electron beam-stabilization-sulphurization approach. *Macromolecular Materials and Engineering, 306*, 2100280.

Furst, K. E., Coyte, R. M., Wood, M., Vengosh, A., & Mitch, W. A. (2019). Disinfection byproducts in Rajasthan, India: Are trihalomethanes a sufficient indicator of disinfection byproduct exposure in low-income countries? *Environmental Science & Technology, 53*(20), 12007–12017. doi:10.1021/acs.est.9b03484.

Ghodrat, M., Rashidi, M., & Samali, B. (2017). Life cycle assessments of incineration treatment for sharp medical waste. In *Energy technology 2017: Carbon dioxide management and other technologies* (pp. 131–143). Cham: Springer International Publishing.

Horikiri, S., Iseki, J., & Minobe, M.. *Process for production of carbon fiber*. https://patents.google.com/patent/US4070446A/en.

Ilyas, S., Srivastava, R. R., & Kim, H. (2020). Disinfection technology and strategies for COVID-19 hospital and bio-medical waste management. *Science of the Total Environment, 749*, 141652. doi:10.1016/j. scitotenv.2020.141652.

Jagatee, S., Priyadarshini, S., Kumbhar, M. M., Tudu, J., & Das, A. P. (2026). An insight into implementing advanced techniques for the removal of emerging micropollutants from wastewater. In A. P. Das, S. Mishra, & M. P. Shah (Eds.), *Decarbonization of Wastewater Pollutants as a Sustainable Solution*. Elsevier https://doi.org/10.1016/B978-0-443-33062-9.00022-9.

Jimoh, J. O., Rahmah, S., Mazelan, S., Jalilah, M., Olasunkanmi, J. B., Lim, L. S., Ghaffar, M. A., Chang, Y. M., Bhubalan, K., & Liew, H. J. (2023). Impact of face mask microplastics pollution on the aquatic environment and aquaculture organisms. *Environmental Pollution (Barking, Essex: 1987), 317*, 120769. doi:10.1016/j.envpol.2022.120769.

Kenny, C., & Priyadarshini, A. (2021). Review of current healthcare waste management methods and their effect on global health. *Healthcare (Basel, Switzerland),, 9*(3), 284. doi:10.3390/healthcare9030284.

Kothari, R., Sahab, S., Singh, H. M., Singh, R. P., Singh, B., Pathania, D., Singh, A., Yadav, S., Allen, T., Singh, S., & Tyagi, V. V. (2021). COVID-19 and waste management in the Indian scenario: Challenges and possible solutions. *Environmental Science and Pollution Research international, 28*(38), 52702–52723. doi:10.1007/s11356-021-15028-5.

Kumbhar, M. M., Sethy, P. S., Swain, S., & Kaur, D. (2024a). Emerging technologies for effective disposal and sustainable management of biomedical waste (BMW) during COVID-19 pandemic. In A. P. Das, & S. Mishra (Eds.), *Impact of COVID-19 waste on environmental pollution and its sustainable management. environmental science and engineering* (pp. 171–195). Springer. doi:10.1007/978-3-031-50840-0_10.

Kumbhar, M. M., Swain, S., Sethy, P. S., & Kaur, D. (2024b). Energy recovery from municipal solid waste through pyrolysis techniques. In M. P. Shah, & A. P. Das (Eds.), *Advanced green technology for environmental sustainability and circular economy* (pp. 149–167). CRC Press. doi:10.1201/9781003517108-9.

Kumbhar, M. M., Tudu, J., Jagatee, S., Priyadarshini, S., & Das, A. P. (2026). Microbial bioremediation of emerging contaminants in aquatic ecosystem for a sustainable environment: Recent strategies and future perspectives. In A. P. Das, S. Mishra, & M. P. Shah (Eds.), *Decarbonization of Wastewater Pollutants as a Sustainable Solution*. Elsevier https://doi.org/10.1016/B978-0-443-33062-9.00018-7.

Kaur, D., Sharan, N., & Kumbhar, M. M. (2026). Marine microplastics: The enduring effects of plastic pollution. In A. P. Das, & S. Dey (Eds.), *Bioremediation as a Greener Solution for Marine Plastic Pollutants*. Elsevier https://doi.org/10.1016/B978-0-443-33182-4.00015-2.

Lee, G., Eui Lee, M., Kim, S.-S., Joh, H.-I., & Lee, S. (2022). Efficient upcycling of polypropylene-based waste disposable masks into hard carbons for anodes in sodium ion batteries. *Journal of Industrial and Engineering Chemistry, 105*, 268–277.

Lindmark, M., Cherukumilli, K., Crider, Y. S., Marcenac, P., Lozier, M., Voth-Gaeddert, L., Lantagne, D. S., Mihelcic, J. R., Zhang, Q. M., Just, C., & Pickering, A. J. (2022). Passive in-line chlorination for drinking water disinfection: A critical review. *Environmental Science & Technology, 56*(13), 9164–9181. doi:10.1021/acs.est.1c08580.

Lopez-Fernandez, M., Jroundi, F., Ruiz-Fresneda, M. A., & Merroun, M. L. (2021). Microbial interaction with and tolerance of radionuclides: Underlying mechanisms and biotechnological applications. *Microbial Biotechnology, 14*(3), 810–828. doi:10.1111/1751-7915.13718.

Manupati, V. K., Ramkumar, M., Baba, V., & Agarwal, A. (2021). Selection of the best healthcare waste disposal techniques during and post-COVID-19 pandemic era. *Journal of Cleaner Production, 281*, 125175. doi:10.1016/j.jclepro.2020.125175.

Manzoor, J., & Sharma, M. (2019). Impact of biomedical waste on the environment and human health. *Environmental Claims Journal, 31*(4), 311–334. doi:10.1080/10406026.2019.1619265.

Mishra, S., Ghosh, S., van Hullebusch, E. D., Singh, S., & Das, A. P. (2023). A critical review on the recovery of base and critical elements from electronic waste-contaminated streams using microbial biotechnology. *Applied Biochemistry and Biotechnology, 195*(12), 7859–7888. doi:10.1007/s12010-023-04440-x.

Ojha, P. C., Satpathy, S. S., Ojha, A. K., Sukla, L. B., & Pradhan, D. (2022). Overcoming challenges due to enhanced biomedical waste generation during COVID-19 pandemic. *The Science of the Total Environment, 832*, 155072. doi:10.1016/j.scitotenv.2022.155072.

Priyadarshini, S., Jagatee, S., Tudu, J., Kumbhar, M. M., & Das, A. P. (2026). Das Microbe-mediated wastewater treatment: Harnessing nature's engineers for sustainable remedies. In A. P. Das, S. Mishra, & M. P. Shah (Eds.), *Decarbonization of Wastewater Pollutants as a Sustainable Solution* (pp. 103–136). Elsevier. https://doi.org/10.1016/B978-0-443-33062-9.00008-4.

Robertson, M., Obando, A. G., Emery, J., & Qiang, Z. (2022). Multifunctional carbon fibers from chemical upcycling of mask waste. *ACS Omega, 7*(14), 12278–12287. doi:10.1021/acsomega.2c00711.

Sharma, S., Zhang, M., Anshika, Gao, J., Zhang, H., & Kota, S. H. (2020). Effect of restricted emissions during COVID-19 on air quality in India. *The Science of the Total Environment, 728*, 138878. doi:10.1016/j.scitotenv.2020.138878.

Silva, A. L. P., Prata, J. C., Walker, T. R., Campos, D., Duarte, A. C., Soares, A. M. V. M., Barcelò, D., & Rocha-Santos, T. (2020). Rethinking and optimising plastic waste management under COVID-19 pandemic: Policy solutions based on redesign and reduction of single-use plastics and personal protective equipment. *The Science of the Total Environment, 742*, 140565. doi:10.1016/j.scitotenv.2020.140565.

Singh, N., Tang, Y., & Ogunseitan, O. A. (2020). Environmentally sustainable management of used personal protective equipment. *Environmental Science & Technology, 54*(14), 8500–8502. doi:10.1021/acs.est.0c03022.

Singh, S., Tom, V., Verma, R., Malik, I., Vashist, M. G., & Dahiya, P. (2022). To study the knowledge about the handling of biomedical waste among health-care workers in a COVID-19 hospital setting. *Journal of Education and Health Promotion, 11*, 193. doi:10.4103/jehp.jehp_871_21.

Subramanian, K. G., Dhanushkodi, M., Satyapriyan, A., Nagarajan, M., & Rajasekaran, P. (2023). Incidence of Pink Water in Chennai, India- An Alarming Bloom. *International Journal of Agriculture Sciences, 15*(1), 12142–12145.

Sussarellu, R., Suquet, M., Thomas, Y., Lambert, C., Fabioux, C., Pernet, M. E., Le Goïc, N., Quillien, V., Mingant, C., Epelboin, Y., Corporeau, C., Guyomarch, J., Robbens, J., Paul-Pont, I., Soudant, P., & Huvet, A. (2016). Oyster reproduction is affected by exposure to polystyrene microplastics. *Proceedings National Academy of Science USA, 113*(9), 2430–2435. doi:10.1073/pnas.1519019113.

Thiounn, T., & Smith, R. C. (2020). Advances and approaches for chemical recycling of plastic waste. *Journal of Polymer Science, 58*(10), 1347–1364.

Usman, I., & Elizabeth, A. Q. (2020). Green operations practice of hospital medical waste management: A case study. *Test Engineering and Management, 83*, 3749–3756. https://www.testmagzine.biz/index.php/testmagzine/article/view/4239.

Tudu, J., Kumbhar, M. M., Priyadarshini, S., Jagatee, S., & Das, A. P. (2026). Research advancements in sustainable wastewater treatment technologies. In A. P. Das, S. Mishra, & M. P. Shah (Eds.), *Decarbonization of Wastewater Pollutants as a Sustainable Solution*. Elsevier https://doi.org/10.1016/B978-0-443-33062-9.00006-0.

WHO (2020a) Shortage of personal protective equipment endangering health workers world-wide. https://www.who.int/news/item/03-03-2020-shortage-of-personal-protectiveequipment-endangering-health-workers-worldwide

WHO (2020b) Q&A on coronaviruses (COVID-19). https://www.who.int/emergencies/diseases/novel-coronavirus-2019/question-and-answers-hub/q-a-detail/q-a-coronaviruses

World Health Organization.. *WHO COVID-19 dashboard data*. https://data.who.int/dashboards/covid19/data.

CHAPTER

Global solutions: A comparative study of regional best practices

7

Harishma Ramesh [a] **and Vinoth Gnana Chellaiyan Devanbu** [b]

[a] *Department of Community Medicine, Tagore Medical College and Hospital, Rathinamangalam, Chennai, Tamil Nadu, India,* [b] *Department of Community Medicine, Chettinad Academy of Research and Education, Chettinad Hospital and Research Institute, Kelambakkam, Tamil Nadu, India*

7.1 Introduction

7.1.1 History

In 1983, the first official documentation on biomedical waste (BMW) management emerged during a meeting organized by the World Health Organization's (WHO) regional office for Europe, which took place in Bergen, Norway (Indian Dental, 2024). The issue of waste management gained significant attention in 1987 when a large volume of hospital waste was found washed up on the East Coast beaches of New Jersey and New York. This infamous event, referred to as the "Syringe Tide" (Sugrue, 1991), led the US Environmental Protection Agency (EPA) to intervene, prompting the US Congress to pass the Medical Waste Tracking Act (MWTA) in 1988 (U.S. Environmental Protection Agency, 2026). The EPA released regulations for medical waste management, guided by the 2-year federal MWTA program. The MWTA came into effect on June 24, 1989, and was enforced in four states: New York, New Jersey, Rhode Island, Puerto Rico, and Connecticut. After collecting data, the EPA concluded that the risk of disease transmission from hospital waste was highest at the point of generation. The MWTA expired in 1991, after which individual states developed their regulations based on the MWTA's recommendations (U.S. Environmental Protection Agency, 2016). This initiative marked the first step toward regulating medical waste management, serving as a foundation for other countries to adopt similar measures. The framework provided by the MWTA set a precedent for global healthcare waste management practices, influencing the development of regulations in various nations aimed at reducing the risks associated with BMW.

The WHO defines medical waste as "*any waste produced through healthcare activities, including items such as used needles and syringes, contaminated dressings, body parts, diagnostic samples, blood, chemicals, pharmaceuticals, medical devices, and radioactive materials*" (World Health Organization, n.d.). Of the total waste

Biomedical Waste Management in the COVID-19 Era: A Global Scenario.
DOI: https://doi.org/10.1016/B978-0-443-21841-5.00007-2

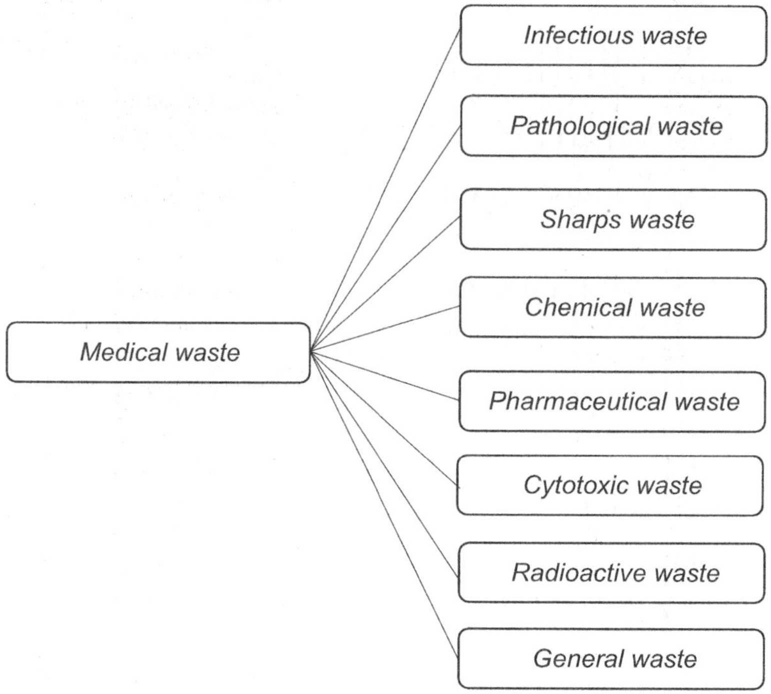

FIGURE 7.1

Categories of medical waste.

produced by healthcare activities, 15% is classified as hazardous, which can be infectious, chemical, or radioactive. The primary sources of health-care waste include:

- Hospitals and health-care facilities
- Laboratories and research centers
- Mortuaries and autopsy centers
- Animal research and testing laboratories
- Blood banks and donation services
- Nursing homes for the elderly (World Health Organization, 2024)

7.1.2 Categories of medical waste

Medical waste is classified into several categories based on its composition and risks to human health and the environment (Fig. 7.1).

7.1.3 Infectious waste

It includes materials such as blood-soaked gauze, diagnostic samples, and cultures of pathogens created during lab work. Additionally, waste from autopsies and specimens from infected animals are also classified here. Items like used bandages and syringes from patients diagnosed with infectious diseases are part of this category. Examples include used surgical gloves contaminated with blood, used syringes and needles, soiled dressings and bandages, contaminated medical equipment such as scalpels, and waste from patients with contagious diseases, like disposable bedding and masks.

7.1.4 Pathological waste

Human tissues, organs, bodily fluids, and body parts require specialized handling due to their potential infection risk comes under this category. It also encompasses contaminated animal carcasses that may harbor diseases transmissible to humans. Examples include amputated limbs from surgical procedures, fetal remains from abortions, removed tumors, infected body fluids such as blood or urine from patients, and pathological specimens that have not yet been analyzed.

7.1.5 Sharps waste

Sharps waste refers to items that can easily puncture or cut, posing a risk of injury and infection. This category includes needles, syringes, and disposable scalpels commonly used in medical settings. Examples of sharps waste include discarded needles from vaccinations, broken glass vials containing hazardous materials, used lancets from blood glucose testing, and surgical blades left over from procedures.

7.1.6 Chemical waste

This waste consists of hazardous substances produced in healthcare and laboratory environments that can pose environmental and health dangers. This includes solvents and reagents used for laboratory testing, as well as disinfectants and sterilizers vital for infection control. Additionally, heavy metals found in medical devices, such as mercury from thermometers, fall into this category. Examples of chemical waste include leftover cleaning agents from a laboratory, expired pharmaceuticals that contain toxic compounds, solvents such as acetone or ethanol used in medical procedures, and discarded chemotherapeutic agents used in cancer treatment.

7.1.7 Pharmaceutical waste

This encompasses expired, unused, or contaminated medications, including vaccines, which must be disposed of correctly to minimize health risks and environmental impact. Examples of pharmaceutical waste include unused antibiotics that have passed their expiration date, discarded chemotherapy drugs that are no longer needed, expired vaccines that cannot be administered, and leftover liquid medications from syringes or vials after a patient's treatment.

7.1.8 Cytotoxic waste

It includes materials that contain genotoxic substances, which may be mutagenic, teratogenic, or carcinogenic, along with cytotoxic drugs and their metabolites used in cancer treatments. Cytotoxic waste includes contaminated personal protective equipment (PPE) used during chemotherapy, unused portions of chemotherapy drugs that have been prepared but not administered, IV bags and tubing that contained cytotoxic drugs after patient treatment, and syringes and needles used to inject cytotoxic medications.

7.1.9 Radioactive waste

This is made up of items contaminated with radionuclides, as well as materials used in radioactive diagnostic procedures and therapies. This type of waste requires specialized methods of storage and disposal to protect public health and the environment from the risks associated with radiation exposure. Radioactive waste includes used diagnostic imaging equipment that contains radioactive materials, contaminated materials such as gloves and syringes from procedures involving radioactive substances, waste generated from the production of radiopharmaceuticals, and discarded sources of radiation used in treatments, such as brachytherapy seeds.

7.1.10 Nonhazardous or general waste

It includes materials that do not pose any biological, chemical, radioactive, or physical hazards, allowing for standard waste disposal methods without special precautions. Nonhazardous or general waste includes items such as empty medication containers that are not considered hazardous, food waste generated in healthcare facilities, used paper and cardboard from administrative offices, and disposable patient care items like clean gowns and bed linens that do not pose an infection risk (World Health Organization, 2024).

7.1.11 Impact of the COVID-19 pandemic on waste generation

The COVID-19 pandemic, which began in 2019, [Coronavirus Disease (COVID-19) Pandemic, 2024], not only strained healthcare systems globally but also significantly increased waste generation due to the surge in medical and protective equipment use. In 2020, the world produced approximately 2.24 billion tons of solid waste, equating to about 0.79 kg per person daily. With rapid urbanization and population growth, global waste production is projected to rise by 73%, reaching 3.88 billion tons by 2050 (World Bank Group, n.d.). This dramatic increase presents a growing challenge for waste management systems worldwide.

As cases surged in 2021, hospital admissions increased. However, some patients were denied due to bed shortages, many individuals self-isolated at home and used prescribed medications, disposing of wrappers and boxes as regular household waste. This resulted in a significant rise in medical waste from households, often improperly handled and treated like everyday trash, further straining the existing waste management systems in all countries (Jindal & Sar, 2023). This pandemic has generated tons of additional medical waste, overwhelming healthcare waste management systems and posing serious risks to human, animal, and environmental health, as highlighted in a recent WHO report. Between March 2020 and November 2021, approximately 87,000 tons of PPE were distributed globally, much of which ended up as waste. Additionally, over 140 million test kits and 8 billion vaccine doses contributed significantly to plastic, chemical, and sharps waste, further intensifying the crisis (Tonnes of, 2024).

7.1.12 Significance of studying regional responses to the crisis

Understanding regional responses to the COVID-19 waste management crisis is crucial for developing effective global solutions. Each region faced unique challenges due to differences in healthcare infrastructure, waste disposal systems, and resource availability (Das et al., 2021). Studying these diverse approaches helps identify the strengths and weaknesses of each strategy, offering valuable insights into how different areas managed the surge in medical waste. By examining regional variations, such as Southeast Asia's (SEARO) focus on monitoring and training (BioSpectrum Asia. (n.d.)) or the Americas' adoption of alternative treatment methods (Mazzei & Specchia, 2023), we can foster knowledge sharing and collaboration. This comparative analysis is essential for improving global healthcare waste management practices and preparing for future public health crises.

7.2 Overview of World Health Organization regions and healthcare waste management

7.2.1 World Health Organization regional structure

The WHO operates through six regional offices, each tasked with addressing the distinct health priorities of its respective member states. These regions include Africa (AFRO), the Americas (PAHO), SEARO, Europe (EURO), the Eastern Mediterranean (EMRO), and the Western Pacific (WPRO). These regional offices play a crucial role in translating the WHO's global health policies into actionable strategies tailored to their local settings. By working closely with individual countries, they tackle public health issues, such as healthcare waste management, while accounting for the unique geographic, socioeconomic, and cultural factors that influence each region. This regional approach ensures that WHO's global initiatives are effectively adapted to meet local challenges (World Health Organization, n.d.). In the following sections, we will examine the responses of each WHO region to the healthcare waste crisis, highlighting one country from each region and its approach to handling medical waste during the pandemic.

7.2.2 Healthcare waste management

The WHO acknowledges the substantial health risks that healthcare waste poses not only to healthcare workers and waste handlers but also to the wider community.

Beyond human health, this waste has profound implications for wildlife and aquatic ecosystems, highlighting the urgent need for effective waste management practices to safeguard both human and environmental health. To address these challenges, the WHO is committed to facilitating and implementing effective healthcare waste management strategies through several key initiatives:

a. *Development of technical guidance materials:* WHO is creating essential resources designed to assess the quantities and types of waste produced in various healthcare facilities. These materials will aid in understanding waste generation patterns and inform better management practices.
b. *Creation of national action plans*: The organization is actively working on formulating national healthcare waste management guidelines and action plans that are customized to meet the specific needs of individual countries. This approach ensures that the strategies are relevant and effective in addressing local challenges.
c. *Building national capacity*: WHO is dedicated to strengthening the management of healthcare waste, particularly in low-income countries, by enhancing national capacity. This focus on capacity building is crucial for ensuring sustainable and effective waste management practices.

Now, we will explore how the six WHO regions addressed the challenges of hospital waste management during the COVID-19 pandemic.

7.3 Southeast Asia region

7.3.1 Response to the increased biological waste during COVID-19

In densely populated, resource-constrained Asian developing countries, the failure to implement effective healthcare waste management systems is still prevalent. Facilities often lack essential practices such as proper waste segregation, safe collection, storage, transportation, and disposal (Abdel-Shafy & Mansour, 2018). Many articles have highlighted critical challenges faced by these nations in managing healthcare waste, noting that recent regulations and policies vary widely in their implementation (Khan et al., 2019). Furthermore, discrepancies in waste generation rates and conflicting measurement methods raise questions about the reliability of data. The absence of waste management training for staff contributes to unsafe handling practices, increasing health risks (Kumar, Somrongthong, & Ahmed, 2016). Additionally, the illegal recycling of hazardous waste and the confusion between landfilling and open dumping exacerbate environmental degradation (Siddiqua, Hahladakis, & Al-Attiya, 2022). To mitigate toxic gas emissions, outdated incineration methods must be replaced with modern techniques like autoclaving, steam sterilization, and pyrolysis (Mazzei & Specchia, 2023). Recognizing the importance of proper healthcare waste management is essential; substantial improvements are necessary to safeguard both human health and the environment in Asian developing countries.

7.3.2 Case study: Country-level innovations—India

According to the report, India generated 28,747.91 tons of BMW from June 2020 to December 2021 (Chand et al., 2021). The significant increase in BMW generation is due to the overwhelming number of patients in hospitals and treatment facilities. This surge posed considerable challenges in efficiently managing the waste, becoming a concern that highlighted existing management deficiencies. Studies revealed that improper waste management practices posed serious risks to healthcare workers' and sanitation staff's health and safety, emphasizing the urgent need for improved protocols (Khosla et al., 2022).

Moreover, there were notable gaps in the entire waste management process, from segregation to collection and disposal, contributing to ineffective management (Singh et al., 2024). Many states in India faced a shortage of common BMW treatment facilities (CBMWTFs), and the restriction of movement across borders made it more difficult (Shefali, 2023). Additionally, neglect and ignorance among various stakeholders exacerbated the issues related to BMW management, underscoring the need for heightened awareness and action to address these critical challenges (Saxena et al., 2021).

To address the significant challenges posed by COVID-19 waste management during the pandemic, the Central Pollution Control Board issued a dedicated guideline specifically for managing COVID-related waste in homes and healthcare centers. These guidelines were presented in an easily understandable pictorial format, facilitating better comprehension and implementation. (Central Pollution Control, 2023). The BMW was effectively managed in hospitals and healthcare centers, with COVID-19-related waste being distinctly identified and safely collected in designated areas (Jindal & Sar, 2023). Both state and central governments have implemented various initiatives, such as setting up quarantine centers, isolation wards, sample collection sites, and laboratories. All stakeholders and existing BMW treatment and disposal facilities were directed to adhere to stringent guidelines for managing waste generated during the diagnosis and treatment of confirmed or suspected COVID-19 cases, building upon established procedures for handling infectious waste (Kaur Mohi et al., 2023).

7.4 Western Pacific region

7.4.1 Response to the increased biological waste during COVID-19

The COVID-19 pandemic also presented significant challenges for the Western Pacific Region's (WPR) BMW management systems, with an estimated 1500 tons of PPE waste from gloves and masks generated to support healthcare needs (World Health Organization, 2022). The surge in waste from COVID-19 testing, vaccinations, and treatment created an urgent need for rapid adaptations in waste management systems across the WPR. Countries such as Cambodia, Mongolia, and the Philippines experienced a significant rise in healthcare waste, with waste generation rates soaring due to pandemic-related activities. The Philippines saw a staggering increase in waste generation, with healthcare waste rising to approximately 634,687.73 metric tons between June 2020 and June 2021. This equates to about 52,890 metric tons of healthcare waste per month, significantly exceeding initial projections. Additionally, the country generated an estimated 8.2 million tons of plastic waste annually during the pandemic, up from 2.7 million tons in 2019 (Apostol et al., 2022).

In response, the region also prioritized improvements in waste treatment infrastructure, focusing on upgrading monitoring systems and adopting low-emission technologies to manage the growing BMW burden better. The existing frameworks across the region were tested by the unprecedented volume of infectious waste. While most countries had general guidelines in place, their preparedness varied. For example, incineration remained the predominant method for treating healthcare waste in several countries, with stringent environmental regulations being applied to minimize emissions (Kenny & Priyadarshini, 2021). However, some countries like the Lao People's Democratic Republic, Mongolia, and Singapore sought to transition to low-carbon technologies and cleaner alternatives to incineration (World Health Organization Regional Office for the Western Pacific, 2022). These shifts were motivated by a growing awareness of the environmental impact of traditional waste treatment methods and were further propelled by the increased waste from the pandemic.

The WPR faced significant challenges in managing healthcare waste, largely due to the absence of comprehensive disaster preparedness protocols. The Pacific health care waste strategy 2013_2015 aimed to address these gaps, but infrastructure deficiencies, insufficient resources, and inadequate treatment facilities hindered effective waste management during the pandemic (Secretariat of the Pacific Regional Environment Programme, 2013). In response, WPR countries focused on upgrading waste treatment infrastructure, implementing enhanced data monitoring, and refining policies to improve waste tracking. Efforts included the adoption of low-emission disposal methods, such as autoclaves and microwave disinfection, to replace traditional incineration, ultimately leading to improved management of the surge in waste generated from high PPE usage and vaccination rollouts (Zhang et al., 2022).

7.4.2 Case study: Country-level innovations—Japan

During the COVID-19 pandemic, Japan confronted significant challenges in managing waste, particularly healthcare waste. As the number of infections rose, there was a notable increase in the disposal of infectious materials, such as used masks and testing kits. Simultaneously, the surge in demand for home deliveries led to an influx of packaging waste, especially cardboard. While waste generation from businesses declined due to temporary closures, the overall rise in household waste raised serious concerns about proper disposal and treatment, exacerbating existing waste management issues, particularly in handling infectious waste (Liu et al., 2021).

Throughout the pandemic, Japan's waste treatment largely consisted of 74% incineration, 17% recycling, and just 3% landfill disposal (Moshkal et al., 2024). In response to these mounting challenges, Japan took decisive measures to reinforce its waste management systems. The Japanese government encouraged a "new way of life," advocating for lifestyle and business adjustments that addressed environmental and energy challenges. In 2020, Japan adopted smart strategies, including the use of Artificial Intelligence (AI) to operate and monitor municipal solid waste incineration plants, enhancing both efficiency and sustainability in waste management practices (Fang et al., 2023). The government prioritized the maintenance and enhancement of waste treatment and recycling infrastructure, which is critical given the country's declining population (Heinrich Böll Stiftung, 2022).

Amid these developments, Japan introduced several innovative solutions to tackle the emerging challenges posed by increased waste and logistical demands. The proposed "multibenefits mobility system" was designed to manage the heightened workload for logistics companies resulting from the surge in home deliveries and waste generation (Koike, 2014). This system features a work chain management system (WCMS) that improves traceability for waste produced by small medical institutions, utilizing smartphones and QR codes to automatically record activity histories. Furthermore, the adoption of remote control and automation technologies, such as Building Energy Management Systems (BEMS) and smart meters, has helped reduce human contact and enhance operational efficiency at waste treatment facilities. Virtual reality (VR) has also emerged as a valuable tool for remote education, particularly in occupational safety training (Onoda, 2020). Also, the ongoing discussions regarding regulatory reforms and the implementation of the "Super City Act" aim to improve data collection and management practices, setting the stage for more efficient waste management systems and enhanced public health safety as Japan adapts to the post-COVID-19 environment (Onoda, 2020).

7.5 African region

7.5.1 Response to the increased biological waste during COVID-19

A 2021 report on waste management estimated that, before the pandemic, Africa generated approximately 282,000 tons of medical waste annually (Chisholm et al., 2021). The African region experienced a significant surge in biological waste, primarily driven by the widespread use of PPE and increased testing and healthcare activities. Studies suggest that 12 billion medical and fabric face masks are discarded monthly, giving the likelihood that an equivalent of about 105,000 tons of face masks per month could be disposed of in the environment by Africans. Also, in general,

15 out of 57 African countries are significant plastic waste contributors, with Nigeria (15%), Ethiopia (8.6%), Egypt (7.6%), DR Congo (6.7%), Tanzania (4.5%), and South Africa (4.4%) (Benson et al., 2021). To address this challenge, regional authorities recommended strengthening healthcare worker training on waste segregation and the proper use of biohazard bags for handling hazardous waste. This was crucial in ensuring that healthcare facilities could effectively manage the influx of waste while minimizing potential health risks to both workers and the community.

Efforts were directed toward sustainable purchasing and encouraging environmentally friendly treatment options, such as autoclaving and microwave treatment, to reduce reliance on incineration, which can pose environmental and health hazards. As African countries received substantial supplies of PPE, the need for enhanced waste segregation and handling practices became increasingly apparent (Oyedotun et al., 2020). Healthcare facilities were supported with biohazard bags and centralized waste treatment options, though many faced ongoing challenges due to limited resources (Organization WH, 2022). Well-designed incinerators and sanitary landfills are advised for safe waste management, yet many countries face gaps in implementing these guidelines. Where infrastructure is lacking, controlled methods like isolated pit burning or safe burial on hospital grounds are preferable to indiscriminate dumping or open burning, which releases harmful emissions (Managing COVID-19 Waste in Africa | WHO | Regional Office for Africa, 2024).

7.5.2 Case study: Country-level innovations—South Africa

In response to the growing BMW challenge, South Africa implemented different solutions to enhance its waste management systems. The South African government recognized the urgent need for better waste segregation practices, like other countries, and the effective disposal of COVID-19 waste (Hantoko et al., 2021). Recognizing the urgency, the government established strict regulations to ensure the safe disposal of PPE and other infectious materials, with guidelines aligned with the Waste Management Act and SANS standards (Department of Environmental Affairs. (n.d.).). One notable initiative was the establishment of public-private partnerships to improve waste management capabilities in healthcare settings. This compliance facilitated the provision of biohazard bags and robust staff training in effective waste segregation practices (Kalu, 2020).

South Africa expanded its waste management infrastructure by investing in centralized waste treatment facilities equipped to handle increased volumes of hazardous waste (Adeleke et al., 2021). Additional efforts included the proactive supply of sharps containers, specifically for vaccine-related waste, which ensured operational readiness for the sharp increase in disposable syringes and needles (Averda, n.d.). The country also implemented strict regulations regarding the disposal of PPE and other infectious waste, mandating that healthcare facilities adopt best practices for waste handling. Moreover, awareness campaigns targeted healthcare workers to emphasize the importance of proper waste management in minimizing health risks and protecting the environment. Drawing from past health outbreak responses, South Africa demonstrated an experienced approach, fostering a collaborative environment in medical waste management with an increased focus on baseline preparedness for potential future surges. These concerted efforts not only improved waste management practices in South Africa but also served as a model for other countries in the African region facing similar challenges (Letho et al., 2021).

7.6 Americas region

7.6.1 Response to the increased biological waste during COVID-19

The COVID-19 pandemic has not only disrupted waste management systems across the Americas but also underscored preexisting vulnerabilities in healthcare infrastructure and socioeconomic disparities. As the continent with the highest COVID-related death toll, early pandemic responses involved strict confinement measures that disrupted recycling activities and significantly increased plastic use. This rapid shift, coupled with a lack of comprehensive waste management data, created obstacles to developing effective strategies to address the surge in waste. Consequently, hospitals, healthcare facilities, and individuals have generated unprecedented volumes of waste, including potentially contaminated masks, gloves, gowns, and other protective equipment (United Nations Environment Programme, 2020).

Additionally, the pandemic highlighted household-level changes in consumption patterns, significantly affecting waste production in Hispanic America, where fragile healthcare and economic systems are coupled with deep-seated inequalities (Leal Filho et al., 2022). In response to the heightened demand for PPE and the associated waste, countries in the region have been compelled to adopt innovative waste management solutions, including improvements in recycling infrastructure and the implementation of nonburning disposal technologies (Martins et al., 2021). These adaptations, however, remain uneven across the region, emphasizing the need for strengthened waste management policies to address both medical and household waste surges during future crises.

7.6.2 Case study: Country-level innovations—Brazil

Until 2019, approximately 61.9% of Brazilian municipalities lacked a selective collection program, resulting in waste being disposed of in sanitary landfills or improperly in dumps and uncontrolled landfills. The situation deteriorated further during the pandemic, as many municipalities with selective waste collection programs halted manual collection and sorting processes. This suspension is likely to reduce public participation in recycling efforts and lead to an increased amount of waste being directed to landfills (Vargas et al., 2024). Brazil, one of the most affected countries in the Americas during the COVID-19 pandemic, faced substantial challenges in managing the increased volumes of healthcare waste generated. In response, the Brazilian government implemented a notable initiative, the establishment of a national policy focused on strengthening waste segregation at healthcare facilities (Anuardo et al., 2022). This included mandatory training programs for healthcare workers on the importance of proper waste handling and segregation practices to minimize risks associated with infectious waste.

Furthermore, Brazil invested in expanding its recycling infrastructure and promoting the use of nonburning disposal technologies, such as autoclaving, to treat BMW safely and sustainably. The government also encouraged partnerships between public and private sectors to improve waste collection and recycling initiatives, aiming to reduce the environmental impact of waste generated during the pandemic. These efforts not only addressed the immediate challenges posed by the surge in waste but also laid the groundwork for a more sustainable waste management framework in Brazil moving forward (Véras e Silva et al., 2023).

7.7 European region

7.7.1 Response to the increased biological waste during COVID-19

The European region faced unprecedented challenges in managing the surge of biomedical and household waste generated by PPE, testing, and vaccinations (Mihai, 2020). On 30 March, the European Center for Disease Prevention and Control (ECDC) released guidelines for infection prevention and control at the household-level, recommending double-bagging waste like used tissues and masks for safe disposal in residual bins (European Centre for Disease Prevention and Control, 2021). To maintain high standards for waste management, the European Commission followed a guideline emphasizing that municipal and medical waste services uphold EU regulations to ensure safe disposal and treatment. Amidst these immediate needs, Zero Waste Europe (ZWE) highlighted the importance of maintaining the EU's long-term circular economy goals. ZWE underscored that despite COVID-19 disruptions, efforts should persist toward the EU's 2020 target of achieving a 50% recycling rate, reinforcing sustainability across the region during the crisis (European Commission. (n.d.)). The focus was on adopting sustainable practices, including the transition to reusable PPE where feasible, which significantly reduced waste volumes. In total, around 4000 tons of PPE waste, predominantly gloves and masks, were generated by healthcare facilities across Europe during the pandemic, necessitating urgent action to minimize the environmental footprint (Rowan & Laffey, 2021). Initiatives emphasized green procurement practices to prioritize recyclable or lower-impact materials alongside the use of nonburning technologies and waste-to-energy solutions for safe waste disposal. These approaches helped to contain the environmental impact of PPE and other COVID-related waste while promoting safer, more sustainable disposal methods. (Organization WH, 2022).

7.7.2 Case study: Country-level innovations—Germany

Germany was recognized as one of the largest producers of packaging waste in the European Union even before the pandemic, generating over 18 million tons each year, which translates to approximately 227.5 kg per person. In contrast, Croatia produces only 67.8 kg per person annually (Süßbauer et al., 2022). In response to the pandemic, the Federal Ministry of the Environment advised households with confirmed or suspected coronavirus cases to dispose of all waste through the residual waste collection stream rather than segregating it as usual. Furthermore, some municipalities have halted the collection of specific waste streams due to staffing shortages, although the requirement to store and collect different waste separately remains in effect (Kumar & Rasquin, 2020).

Germany's approach to managing increased BMW during the COVID-19 pandemic provides an exemplary model of effective waste management practices. The country's strategy centered around reusable PPE for healthcare workers and stringent recycling protocols for single-use items. For instance, medical masks and gowns made from recyclable materials were prioritized, and waste treatment facilities adapted their processes to handle the influx of COVID-19-related waste. Additionally, Germany invested in waste-to-energy infrastructure, which converted medical waste into usable energy. This method not only reduced the overall waste volume but also contributed to the country's renewable energy supply, showcasing a circular economy model. Germany's approach highlights the importance of infrastructure investment and policy support to manage and repurpose waste effectively (Awasthi et al., 2019).

7.8 Eastern mediterranean region

7.8.1 Response to the increased biological waste during COVID-19

The high usage of PPE for both healthcare workers and the general public, combined with the increase in medical waste from testing and treatment facilities, overwhelmed many waste management infrastructures. In several areas, improper disposal practices, such as open burning or dumping in poorly managed landfills, pose significant health and environmental risks, especially in lower-income countries where waste collection and treatment resources are limited. This mismanagement threatens neighboring communities and endangers waste handlers and street sweepers who come in direct contact with improperly discarded items. (Abubakar et al., 2022). In response, WHO and national bodies have advocated for safer disposal practices and issued guidance to ensure that infectious and noninfectious waste are properly segregated and managed according to their risk levels (World Health Organization, 2024).

The region's response included emergency segregation protocols and increased reliance on incineration, autoclaving, and chemical disinfection as preferred waste treatment methods. WHO's Regional Center for Environmental Health Action provided technical support to help countries safely manage COVID-19 waste, highlighting the importance of capacity building and occupational safety for waste handlers. Countries like Jordan implemented initiatives to prevent PPE from re-entering circulation by ensuring secure disposal sites. Additional WHO recommendations included providing strong bags for community-disposed PPE and enforcing strict regulations to prevent PPE reselling. While these measures helped to contain the spread of COVID-19 through waste, the need for sustainable and resilient waste management practices remains critical in the Eastern Mediterranean Region (EMR) (Al-Emam & Al-Yousfi, 2021).

Approximately 9000 tons of PPE-related waste, including masks and gloves, were generated in the region as a result of healthcare response efforts and global health distributions (Organization WH, 2022). To manage the surge, EMR countries implemented emergency waste segregation protocols, ensuring that infectious waste was separated from general waste. Many facilities adopted incineration as the primary method for disposing of COVID-19-related waste, while others explored alternative, environmentally sustainable disposal methods like chemical disinfection and autoclaving.

7.8.2 Case study: Country-level innovations—Jordan

The COVID-19 pandemic has significantly affected various sectors worldwide, including medical waste management, an area that has received comparatively less attention, particularly in Jordan. Studies revealed that government measures, such as curfews, greatly influenced medical waste quantities. While hospitals treating COVID-19 patients, private hospitals, and institutions experienced a threefold increase in waste generation during 2020 compared to 2018, government hospitals, healthcare centers, laboratories, and industries reported a decline in waste production, with some generating zero waste during the curfew period. (Alzboon et al., 2022). These results highlight the need for enhanced monitoring and control measures to prevent illegal disposal and ensure proper handling and transportation of medical waste. Jordan's response to the surge in COVID-19 BMW serves as an exemplary case within the EMR. With the sharp increase in infectious waste from PPE, testing, and vaccinations, Jordan implemented robust emergency segregation and treatment protocols to ensure the safe disposal of hazardous waste (Abdelnabi, 2023).

In response to the escalating BMW challenge, Jordan implemented innovative solutions to manage the surge in waste generated from PPE, testing, and vaccinations. The country adopted stringent emergency segregation and treatment protocols for hazardous waste, prioritizing incineration for high-risk materials while focusing on advanced technologies to minimize environmental impact. Additionally, Jordan invested in autoclaving technology for nonhazardous COVID-19 waste, facilitating sustainable treatment methods. This dual approach allowed Jordan to effectively manage the increased waste load while preventing environmental pollution. The experience of Jordan underscores the critical importance of adaptable infrastructure and emergency preparedness in healthcare waste management, serving as a valuable example for other countries in the EMR facing similar challenges (United Nations Development, 2024).

7.8.3 Cross-regional insights and lessons learned

The COVID-19 pandemic exposed significant gaps in healthcare waste management across regions, with factors such as insufficient infrastructure, lack of awareness, and inadequate enforcement of regulations contributing to the challenges. While some countries managed to adapt quickly, many lacked established waste management systems capable of handling the sharp increase in BMW. Key reasons for these difficulties include limited training on the health hazards posed by healthcare waste, lack of financial and human resources, and insufficient prioritization of waste management in the healthcare sector. The absence of effective regulations and enforcement further compounded these issues, as did the lack of structured disposal systems. Each WHO region faced distinct challenges and responded uniquely based on available resources, environmental constraints, and government initiatives. The following insights highlight regional responses and lessons learned, along with targeted recommendations for future resilience.

7.9 Recommendations for improvement

7.9.1 Southeast Asia region

1. *Increase awareness and training on waste management*: There is a need to improve understanding among healthcare workers and the general public about the types and risks associated with COVID-19 waste. Training on proper waste segregation and disposal practices should be incorporated into routine healthcare practices.
2. *Enhance segregation protocols and expand CBMWTFs*: Establishing strict protocols for segregating waste at the source is critical to prevent health risks. Additionally, increasing both the number and capacity of CBMWTFs is essential to managing waste from healthcare facilities, especially in densely populated urban areas.
3. *Implement innovative recycling methods*: Promoting sustainable waste disposal methods, including recycling and resource recovery from noninfectious waste, can help reduce the overall waste burden. Local governments can encourage this by providing incentives for private sector participation in recycling initiatives.
4. *Develop and enforce stronger policies*: Adopting robust waste management policies with dedicated budget allocations and periodic audits will ensure that healthcare facilities adhere to safe waste disposal practices.

7.9.2 Western Pacific region

1. *Expand waste treatment and recycling infrastructure*: Significant investment is required to build an infrastructure capable of managing both healthcare and household waste. This includes advanced technologies for the safe processing of infectious waste and recycling facilities to manage the increased volume sustainably.

2. *Increase public awareness campaigns*: Education campaigns about safe waste disposal practices, particularly for PPE, can help reduce health risks and environmental contamination. Public transparency and communication can foster trust, encourage compliance, and reduce resistance to waste management practices.
3. *Foster multisector collaboration*: Improved collaboration among government agencies, private sector stakeholders, and community organizations is crucial for efficient resource allocation, better data sharing, and enhanced waste management practices. Grassroots recycling programs can further alleviate the waste burden.
4. *Leverage technology in waste management*: Adopting technological advancements such as AI and big data analytics for tracking waste and optimizing resource allocation can streamline operations and improve efficiency across the region.

7.9.3 African region

1. *Comprehensive training programs*: Training healthcare workers on effective waste segregation and handling practices is essential to minimize infection risks. This can be achieved through regular workshops and integration of waste management training into healthcare curricula.
2. *Expand waste management infrastructure*: Investing in centralized waste treatment facilities with environmentally friendly technologies is necessary to manage infectious waste efficiently. Autoclaves and microwaves can provide sustainable alternatives to incineration.
3. *Promote cross-border collaboration*: Fostering partnerships across African nations can enhance resource sharing, optimize infrastructure, and create opportunities for shared best practices in waste management.
4. *Enhance public education*: Increasing public awareness of the importance of proper waste disposal and its environmental impact will build community support for waste management initiatives and promote safer practices.

7.9.4 Americas region

1. *Improve recycling infrastructure*: Expanding recycling capabilities is crucial to managing the high volumes of PPE and other waste sustainably. Recycling facilities and policies that incentivize recycling can help reduce environmental impacts.
2. *Encourage nonburning disposal methods*: Emphasizing nonburning methods such as autoclaving over traditional incineration can help minimize harmful emissions, protecting both human health and the environment.
3. *Optimize reverse logistics*: Implementing reverse logistics in waste management allows for safe and efficient disposal of infectious waste from healthcare facilities, ensuring compliance with safety standards and reducing the risk of contamination.
4. *Promote public engagement and awareness*: Educating communities on proper waste handling, recycling practices, and environmental impacts can increase community involvement in sustainable waste management initiatives.

7.9.5 European region

1. *Strengthen reusable PPE policies*: Expanding policies and funding for reusable PPE production can significantly reduce waste. Setting procurement standards for reusable PPE across Europe could lead to long-term sustainability.
2. *Develop recycling and waste-to-energy infrastructure*: Investing in recycling systems and waste-to-energy facilities can help reduce environmental harm and transform waste into a resource, aligning with Europe's circular economy goals.
3. *Establish standardized monitoring systems*: Routine monitoring and data collection on waste management practices should be standardized across Europe to ensure consistency and inform policy adjustments.
4. *Promote eco-friendly procurement*: Green procurement policies that prioritize recyclable or biodegradable materials can reduce future waste generation and support Europe's leadership in sustainable healthcare waste management.

7.9.6 Eastern Mediterranean region

1. *Enhance sustainable waste treatment options*: Increased investment in alternative waste treatment technologies, such as autoclaving and microwaving, can provide environmentally friendly disposal options and reduce reliance on incineration.
2. *Implement permanent waste segregation policies*: Expanding emergency protocols into formal policies will improve waste management during future health crises and standardize practices across healthcare facilities.
3. *Encourage regional collaboration*: Sharing resources, best practices, and waste treatment technology across EMR countries can optimize resource use and support effective waste management responses.
4. *Invest in education and training*: Regular training for healthcare workers on waste segregation and safe disposal practices can strengthen the region's capacity to manage waste effectively and safely.

These cross-regional insights reveal the diverse approaches taken by WHO regions to manage healthcare waste challenges during the COVID-19 pandemic. While each region faced unique challenges and implemented distinct strategies, common themes include the need for sustainable waste treatment infrastructure, training and awareness programs, and strengthened policy frameworks to support resilient and environmentally friendly waste management practices.

7.10 Global policy recommendations

Effective management of healthcare waste is critical to prevent health risks associated with improper practices, such as exposure to infectious agents and toxic substances. Key strategies for enhancing healthcare waste management include reducing waste generation and ensuring proper segregation. Establishing robust systems, alongside stringent oversight and regulation, is essential for improving waste segregation, treatment, and disposal practices to meet national and international standards.

Wherever possible, prioritizing safe and environmentally sound treatment methods for hazardous healthcare waste—such as autoclaving, microwaving, steam treatment with internal mixing, and chemical treatment—is vital, rather than relying solely on incineration. A comprehensive system that clearly defines responsibilities, allocates resources, and addresses waste handling and disposal is crucial for long-term improvement. Raising awareness about the risks of healthcare waste and promoting safe practices is essential. Ultimately, adopting environmentally friendly management options will protect individuals during the collection, handling, storage, transportation, treatment, and disposal of waste.

7.11 Conclusion

The WHO outlines key guidelines for healthcare waste management, which include reducing unnecessary waste, separating general from hazardous waste, and treating waste to protect communities and healthcare workers. WHO recommends alternatives to incineration, such as autoclaving or steam treatment, and stresses that waste bags and sharps containers should not be overfilled and must be sealed appropriately. Collection times should match waste production levels, with accessible replacement containers at each site. Education, training, and visual reminders for healthcare workers on waste management protocols are essential. Additionally, WHO supports technical guidance, national policy development, and capacity building, especially in low-income countries.

Abbreviations

AFRO	Africa
PAHO	Americas
AI	Artificial Intelligence
BMW	Biomedical Waste
BEMS	Building Energy Management Systems
CBMWTFs	Common Biomedical Waste Treatment Facilities
EMRO	Eastern Mediterranean
EMR	Eastern Mediterranean Region
EPA	Environmental Protection Agency
EURO	Europe
ECDC	European Centre for Disease Prevention and Control
MWTA	Medical Waste Tracking Act
PPE	Personal Protective Equipment
SEARO	Southeast Asia
VR	Virtual Reality

WPRO	Western Pacific
WPR	Western Pacific Region
WCMS	Work Chain Management System
WHO	World Health Organization
ZWE	Zero Waste Europe

References

Abdelnabi, Z.H. (2023). Knowledge, attitudes, and practices of healthcare workers regarding medical waste management during the COVID-19 era: The case of the Southern West Bank [Master's thesis, Hebron University]. Hebron University DSpace Repository. http://dspace.hebron.edu:8080/xmlui/handle/123456789/1312.

Abdel-Shafy, H. I., & Mansour, M. S. M. (2018). Solid waste issue: Sources, composition, disposal, recycling, and valorization. *Egyptian Journal of Petroleum, 27*(4), 1275–1290. https://www.sciencedirect.com/science/article/pii/S1110062118301375.

Abubakar, I. R., Maniruzzaman, K. M., Dano, U. L., AlShihri, F. S., AlShammari, M. S., Ahmed, S. M. S., ... Alrawaf, T. I. (2022). Environmental sustainability impacts of solid waste management practices in the Global South. *International Journal of Environmental Research and Public Health, 19*(19), 12717. https://doi.org/10.3390/ijerph191912717.

Adeleke, O., Akinlabi, S., Jen, T. C., & Dunmade, I. (2021). Towards sustainability in municipal solid waste management in South Africa: A survey of challenges and prospects. *Transactions of the Royal Society of South Africa, 76*(1), 53–66. https://doi.org/10.1080/0035919X.2020.1858366.

Al-Emam, R., & Al-Yousfi, B. (2021). Waste management of used personal protective equipment during the COVID-19 pandemic in the Eastern Mediterranean Region. *Eastern Mediterranean Health Journal, 27*(11), 1034–1035. https://doi.org/10.26719/2021.27.11.1034.

Alzboon, K., Shawaqfeh, M., Shawaqfah, M., Ammary, B., & Almomani, F. (2022). Impact of COVID-19 on medical waste management in Jordan. *Pollution Research*, 853–859.

Averda. (n.d.). South Africa should expect a surge in medical waste during third wave. Retrieved July 9, 2026, from https://www.averda.com/rsa/news/south-africa-should-expect-a-surge-in-medical-waste-during-third-wave.

Anuardo, R. G., Espuny, M., Costa, A. C. F., & Oliveira, O. J. (2022). Toward a cleaner and more sustainable world: A framework to develop and improve waste management through organizations, governments, and academia. *Heliyon, 8*(4). https://www.cell.com/heliyon/abstract/S2405-8440(22)00513-8.

Apostol, G. L. C., Acolola, A. G. A., Edillon, M. A., & Valenzuela, S. (2022). How comprehensive and effective are waste management policies during the COVID-19 pandemic? *Perspectives From the Philippines. Front Public Health, 10*, 958241. https://pmc.ncbi.nlm.nih.gov/articles/PMC9432801/.

Awasthi, M. K., Sarsaiya, S., Chen, H., Wang, Q., Wang, M., Awasthi, S. K., ... Zhang, Z. (2019). Global status of waste-to-energy technology. In S. Kumar, R. Kumar, & A. Pandey (Eds.), *Current developments in biotechnology and bioengineering: Waste treatment processes for energy generation* (pp. 31–52). Elsevier. https://doi.org/10.1016/B978-0-444-64083-3.00003-8.

Benson, N. U., OH, F.-A., Bassey, D. E., & Atayero, A. A. (2021). COVID-19 pandemic and emerging plastic-based personal protective equipment waste pollution and management in Africa. *Journal of Environmental Chemical Engineering, 9*(3), 105222. https://www.sciencedirect.com/science/article/pii/S2213343721001998.

BioSpectrum Asia. (n.d.). APAC lays focus on medical waste management https://www.biospectrumasia.com/opinion/29/18824/apac-lays-focus-on-medical-waste-management.html.

Chand, S., Shastry, C. S., Hiremath, S., Joel, J. J., Krishnabhat, C. H., & Mateti, U. V. (2021). Updates on biomedical waste management during COVID-19: The Indian scenario. *Clinical Epidemiology and Global Health, 11*, 100715.

Central Pollution Control, Board.. *COVID-19 waste management.* https://cpcb.nic.in/covid-waste-management/.

Coronavirus disease (COVID-19) pandemic. (2024). https://www.who.int/europe/emergencies/situations/covid-19.

Chisholm, J. M., Zamani, R., Negm, A. M., Said, N., Abdel-Daiem, M. M., Dibaj, M., & Akrami, M. (2021). Sustainable waste management of medical waste in African developing countries: A narrative review. *Waste Management & Research, 39*(9), 1149–1163. https://doi.org/10.1177/0734242X211029175.

Department of Environmental Affairs. Republic of South Africa. (n.d.). National Waste Management Strategy. http://wastepolicy.environment.gov.za/book/export/html/197. (Accessed 9 July 2026).

European Centre for Disease Prevention and Control.. *February 9). Infection prevention and control and preparedness for COVID-19 in healthcare settings: Sixth update.* https://www.ecdc.europa.eu/en/publications-data/infection-prevention-and-control-and-preparedness-covid-19-healthcare-settings.

Das, A. K, Islam, M. N., Billah, M. M., & Sarker, A. (2021). COVID-19 pandemic and healthcare solid waste management strategy – A mini-review. *Science of the Total Environment, 778*, 146220. https://www.sciencedirect.com/science/article/pii/S0048969721012882.

European Commission. (n.d.).. Waste and recycling. https://environment.ec.europa.eu/topics/waste-and-recycling_en. (Accessed 9 July 2026).

Fang, B., Yu, J., Chen, Z., Osman, A. I., Farghali, M., Ihara, I., ... Yap, P.-S. (2023). Artificial intelligence for waste management in smart cities: A review. *Environmental Chemistry Letters, 21*, 1959–1989. https://doi.org/10.1007/s10311-023-01604-3.

Heinrich Böll Stiftung Hong Kong. (2022). Plastic Atlas Asia 2022. Retrieved July 9, 2026, from https://hk.boell.org/sites/default/files/2022-05/PlasticAtlasAsia2022_en_WEB_0.pdf.

Hantoko, D., Li, X., Pariatamby, A., Yoshikawa, K., Horttanainen, M., & Yan, M. (2021). Challenges and practices on waste management and disposal during COVID-19 pandemic. *Journal of Environmental Management, 286*, 112140. https://www.sciencedirect.com/science/article/pii/S0301479721002024.

Indian Dental, Association. (2024). *About us* https://www.ida.org.in/Membership/Details/History.

Jindal, M. K., & Sar, S. K. (2023). Medical waste management during COVID-19 situation in India: Perspective towards safe environment. *Waste Management Bulletin, 1*(1), 1. https://pmc.ncbi.nlm.nih.gov/articles/PMC10203865/.

Kalu, K., & Falola, T. (2020). *Reflections on leadership and institutions in Africa* (p. 263). Rowman & Littlefield.

Kaur Mohi, G., Yadav, B., & Bora, I. (2023). Biomedical waste management during COVID-19 pandemic: A medical institute perspective. *International Journal of Health*

Sciences and Research, 13(2), 98–104. https://www.ijhsr.org/IJHSR_Vol.13_Issue.2_Feb2023/IJHSR16.pdf.

Kenny, C., & Priyadarshini, A. (2021). Review of current healthcare waste management methods and their effect on global health. *Healthcare, 9*(3), 284. https://www.mdpi.com/2227-9032/9/3/284.

Khan, B. A., Cheng, L., Khan, A. A., & Ahmed, H. (2019). Healthcare waste management in Asian developing countries: A mini review. *Waste Management & Research: The Journal for a Sustainable Circular Economy, 37*(9), 863–875. https://doi.org/10.1177/0734242X19857470.

Khosla, R., Jha, A., Dua, S., Varmani, S. G., Rajput, N., & Pani, B. (2022). Upsurge in biomedical waste due to COVID-19 in India: A statistical correlation, challenges and recommendations. *Frontiers in Environmental Science, 10*. https://www.frontiersin.org/journals/environmental-science/articles/10.3389/fenvs.2022.1022098/full.

Koike, H. (2014). Mobility perspective for a local city in Japan. *IATSS Research, 38*(1), 32–39. https://www.sciencedirect.com/science/article/pii/S0386111214000168.

Kumar, R., Somrongthong, R., & Ahmed, J. (2016). Impact of waste management training intervention on knowledge, attitude and practices of teaching hospital workers in Pakistan. *Pakistan Journal of Medical Sciences, 32*(3), 705–710. https://doi.org/10.12669/pjms.323.9903.

Kumar, P., & Rasquin, C. (2020). The coronavirus crisis and waste management in Germany. Research Institute for Sustainability at GFZ. Retrieved July 9, 2026, from https://www.rifs-potsdam.de/en/blog/2020/04/coronavirus-crisis-and-waste-management-germany.

Leal Filho, W., Lange Salvia, A., Sierra, J., Fletcher, C. A., Banks, C. E., Velazquez, L., … Neiva, S. (2022). COVID-19 and households waste in Hispanic America: An assessment of trends. *Sustainability, 14*(24), 16552. https://doi.org/10.3390/su142416552.

Letho, Z., Yangdon, T., Lhamo, C., Limbu, C. B., Yoezer, S., Jamtsho, T., … Tshering, D. (2021). Awareness and practice of medical waste management among healthcare providers in National Referral Hospital. *PLOS ONE, 16*(1), e0243817. https://doi.org/10.1371/journal.pone.0243817.

Liu, X., Yu, J., Okubo, K., Sato, M., & Aoki, T. (2021). Case study on the efficiency of recycling companies' waste paper collection stations in Japan. *Sustainability, 13*(20), 11536. https://www.mdpi.com/2071-1050/13/20/11536.

Managing COVID-19 waste in Africa | WHO | regional office for Africa. (2024). https://www.afro.who.int/news/managing-covid-19-waste-africa.

Mazzei, H. G., & Specchia, S. (2023). Latest insights on technologies for the treatment of solid medical waste: A review. *Journal of Environmental Chemical Engineering, 11*(2), 109309. https://www.sciencedirect.com/science/article/pii/S2213343723000489.

Martins, M.A., Pataca, L.C.M., Santos, E. de S., Faria, S.M. da C., Cruz, L.A., Silva, G.A.S., & Mol, M.P.G. (2021). Generation of infectious waste during the COVID-19 pandemic: The case of a Brazilian hospital. Waste Management & Research, 39(10), 1245–1255. https://doi.org/10.1177/0734242X211049301.

Mihai, F. C. (2020). Assessment of COVID-19 waste flows during the emergency state in romania and related public health and environmental concerns. *International Journal of Environmental Research and Public Health, 17*(15), 5439. https://www.mdpi.com/1660-4601/17/15/5439.

Moshkal, M., Akhapov, Y., & Ogihara, A. (2024). Sustainable waste management in Japan: Challenges, achievements, and future prospects: A review. *Sustainability, 16*(17), 7347. https://www.mdpi.com/2071-1050/16/17/7347.

Onoda, H. (2020). Smart approaches to waste management for post-COVID-19 smart cities in Japan. *IET Smart Cities, 2*(2), 89–94. https://doi.org/10.1049/iet-smc.2020.0051.

Organization WH. Global analysis of healthcare waste in the context of COVID-19: status, impacts and recommendations. World Health Organization; 2022. 71 p.

Oyedotun, T. D. T., Kasim, O. F., Famewo, A., Oyedotun, T. D., Moonsammy, S., Ally, N., & Renn-Moonsammy, D.-M. (2020). Municipal waste management in the era of COVID-19: Perceptions, practices, and potentials for research in developing countries. *Research in Globalization, 2*, 100033. https://doi.org/10.1016/j.resglo.2020.100033.

Rowan, N. J., & Laffey, J. G. (2021). Unlocking the surge in demand for personal and protective equipment (PPE) and improvised face coverings arising from coronavirus disease (COVID-19) pandemic – Implications for efficacy, re-use and sustainable waste management. *Science of the Total Environment, 752*, 142259. https://www.sciencedirect.com/science/article/pii/S0048969720357880.

Saxena, P., Pradhan, I. P., & Kumar, D. (2021). Redefining bio medical waste management during COVID- 19 in India: A way forward. *Materials Today: Proceedings, 60*, 849. https://pmc.ncbi.nlm.nih.gov/articles/PMC8506110/.

Shefali (2023). Generation, types, and impacts of biomedical waste during COVID-19: Indian context. *Environmental Analysis, Health and Toxicology, 38*, e2023019. https://pmc.ncbi.nlm.nih.gov/articles/PMC10834080/.

Siddiqua, A., Hahladakis, J. N., & Al-Attiya, W. A. K. A. (2022). An overview of the environmental pollution and health effects associated with waste landfilling and open dumping. *Environmental Science and Pollution Research, 29*(39), 58514–58536. https://doi.org/10.1007/s11356-022-21578-z.

Singh, M., Singh, M., & Singh, S. K. (2024). Tackling municipal solid waste crisis in India: Insights into cutting-edge technologies and risk assessment. *Science of the Total Environment, 917*, 170453. https://www.sciencedirect.com/science/article/pii/S0048969724005904.

Sugrue, D. E. (1991). Protecting our surf from syringes: The comprehensive regulated medical waste management act. *Seton Hall Legislative Journal, 15*, 568. https://heinonline.org/HOL/Page?handle=hein.journals/sethlegj15&id=578&div=&collection=.

Secretariat of the Pacific Regional Environment Programme. (2013). Pacific health care waste: A regional strategy and action plan 2013–2015. Retrieved July 9, 2026, from https://www.sprep.org/attachments/Pacific_Medical_Waste_Strategy_2013(11)_with_wmark_(2).pdf.

Süßbauer, E., Wilts, H., Otto, S.J., Schinkel, J., Wenzel, K., Dehning, R.-L., & Caspers, J. (2022). Way out of the one-way? Effects of the COVID-19 pandemic on the generation of waste from packaging in Germany. Sustainability Management Forum, 30(1–4), 1–9. https://doi.org/10.1007/s00550-022-00525-z.

Tonnes of (2024). *COVID-19 health care waste expose urgent need to improve waste management systems* https://www.who.int/news/item/01-02-2022-tonnes-of-covid-19-health-care-waste-expose-urgent-need-to-improve-waste-management-systems.

United Nations Development, Programme.. *How UNDP supports Jordan handling medical waste amid pandemic.* https://www.undp.org/jordan/press-releases/how-undp-supports-jordan-handling-medical-waste-amid-pandemic.

U.S. Environmental Protection Agency. (2026). Medical waste. https://www.epa.gov/rcra/medical-waste. (Accessed 9 July 2026).

U.S. Environmental Protection Agency. (2016). Medical Waste Tracking Act of.

United Nations Environment Programme.. *COVID-19 waste management factsheets*. https://www.unep.org/resources/factsheet/covid-19-waste-management-factsheets.

Vargas, I. S., Gomes, F. B. R., & Castro, S. R. (2024). Reflections on COVID-19 pandemic and waste management in developing countries: A case study in São Paulo city, Brazil. *Waste Management & Research: The Journal for a Sustainable Circular Economy, 42*(1), 41–50. https://journals.sagepub.com/doi/10.1177/0734242X231175817.

Véras e Silva, A. RM. ã., de Freitas Iwata, B., dos Santos, M. DS. F., & Neto, J. M. M. (2023). Impacts and regulations of healthcare solid waste management during the COVID-19 pandemic: A systematic review. *Sustainability, 15*(19), 14462. https://www.mdpi.com/2071-1050/15/19/14462.

World Health Organization.. *Global analysis of health care waste in the context of COVID-19: Status, impacts and recommendations*. https://www.who.int/publications/i/item/9789240039612.

World Health Organization. (2024). Health-care waste. https://www.who.int/news-room/fact-sheets/detail/health-care-waste.

World Bank Group. (n.d.). What a Waste 3.0: Global snapshot of solid waste management toward circularity until 2050. Retrieved July 9, 2026, from https://www.worldbank.org/en/topic/urbandevelopment/brief/solid-waste-management.

World Health Organization, n.d. Medical waste. Available at: https://www.who.int/india/health-topics/medical-waste (accessed 9 July 2026).

World Health Organization. (n.d.). WHO regional offices. Retrieved July 9, 2026, from https://www.who.int/about/who-we-are/regional-offices.

World Health Organization Regional Office for the Western Pacific. (2022). *Annual report 2021: WHO Asia-Pacific Centre for Environment and Health in the Western Pacific Region (WPR-2022-DPM-001)* https://www.who.int/publications/i/item/WPR-2022-DPM-001.

Zhang, X., Liu, C., Chen, Y., Zheng, G., & Chen, Y. (2022). Source separation, transportation, pretreatment, and valorization of municipal solid waste: A critical review. *Environment, Development and Sustainability, 24*(10), 11471–11513. https://doi.org/10.1007/s10668-021-01932-w.

CHAPTER

8

Biomedical hazards: Uncovering the environmental and public health threats of biomedical waste

Henry Sawyerr [a] **and Edet Otto** [b]

[a] *Department of Environmental Health, Kwara State University, Malete, Nigeria,* [b] *Department of Environmental Health, Ajayi Crowther University, Oyo, Nigeria*

8.1 Introduction

Properly managing biomedical waste (BMW) is a significant global environmental and public health concern, especially in economies with limited resources. The magnitude of the grim report on BMW justifies the seriousness and global concern surrounding the situation. For instance, it is estimated that globally, at least 5.2 million deaths occur annually, including 4 million children, because of illnesses caused by unmanaged BMW (Rahman et al., 2020), linked with the rapid increase of healthcare facilities and excessive use of medical facilities resulting in increased waste generation, improper disposal, and illegal recycling practices (Pachauri et al., 2019; Sharma et al., 2020). The current situation highlights the importance of implementing an efficient waste management system to effectively manage the growing amount of BMW, considering the associated health and environmental risks.

However, the current scenario has become more alarming following the COVID-19 outbreak, with increased demand for medical resources leading to a significant rise in BMW generation and management, which has major implications for health and the environment. For example, studies show that there has been a sharp rise in the demand, use, and generation of a substantial amount of BMWs, causing considerable pressure on waste management systems globally (Singh et al., 2020a; You et al., 2020), particularly in developing countries where the implementation of safe waste disposal methods is insufficient (Wise, 2022). Similarly, evidence suggests that BMW experienced a significant increase, ranging from 350% to 500%, in several countries

Biomedical Waste Management in the COVID-19 Era: A Global Scenario.
DOI: https://doi.org/10.1016/B978-0-443-21841-5.00005-9

following the COVID-19 pandemic (Fadaei, 2021). As such, significant environmental effects have been well-documented, including soil and water contamination, as well as detrimental impacts on wildlife and ecosystems (Ansari et al., 2019; Andeobu et al., 2022; Parida et al., 2022), including potential dangers to healthcare workers (HCWs), waste handlers, and the public through potential exposure to infectious agents and harmful chemicals (Javid & Sharma, 2019; Mitiku et al., 2022; Ojha et al., 2022), emphasizing the need for a better understanding of the issues and challenges surrounding BMW's handling, as well as their implications for environmental and public health.

Therefore, this chapter aims to provide a detailed overview of BMW issues, covering their characteristics and categories, including an analysis of the environmental and health impacts associated with improper disposal of BMW, focusing on recent studies addressing these issues, particularly in light of the COVID-19 pandemic. The challenges of BMW management are explored, highlighting the pressing need for improved and sustainable practices. Additionally, global best practices are emphasized, and innovative solutions are proposed to enhance the efficiency and sustainability of BMW handling. This is crucial not only for reducing the current effects of the pandemic but also for strengthening our readiness for future public health emergencies. Thus, this chapter offers valuable insight for promoting a safer, healthier, and more sustainable future, highlighting the importance of applying the experiences from the COVID-19 pandemic to drive long-lasting improvements in BMW management.

8.2 Biomedical waste: Concept and definition

The interpretation of BMW differs across nations and academic and scientific communities. This variation depends on perception, resulting in different connotations such as hospital waste (Chandrappa & Das, 2012), medical waste (Aymen & Firas, 2018), clinical waste (Khan et al., 2019a; Yu et al., 2020), or even healthcare waste (Ansari et al., 2019). The lack of a universally accepted definition for BMW has resulted in a lack of consistency in how BMW is categorized and disposed of (Insa et al., 2010). As a result, it becomes difficult to compare data on BMW across different countries (Bendjoudi et al., 2009; Nemathaga et al., 2008). However, in basic terms, BMW typically refers to a specific kind of waste produced in healthcare facilities, such as hospitals, during medical procedures, diagnoses, treatments, or vaccinations for humans or animals, as well as during research or the manufacturing and testing of biological products. More so, as expanded by WHO and discussed in Windfeld & Brooks' Study (2015), the definition now encompasses waste produced from providing healthcare services in people's homes. According to Dehghani et al. (2019), BMWs make up around 1%–2% of the overall urban waste, which adds a significant amount of waste material to the general municipal solid waste stream, with potential adverse effects on the environment and public health, regardless of the type of waste generated. This is particularly concerning in areas where proper management of BMWs is lacking.

8.3 Categorization and characteristics of biomedical waste

The general perception is that BMWs are hazardous due to the activities carried out in those settings. Nevertheless, studies indicate that around 85% of waste resulting from healthcare activities is nonhazardous. In comparison, the remaining 15% is considered hazardous, with potential harm due to being infectious or radioactive (Janik-Karpinska, 2023), suggesting that BMW includes both dangerous and non-dangerous waste materials, which could potentially endanger public health and the environment if not correctly handled. Hence, there is a need to classify BMW based on proper handling and safety considerations. On this basis, BMW is categorized into different classes based on source, type, and inherent risk. These categories include infectious waste, pathological waste, sharp waste, pharmaceutical waste, chemical waste, cytotoxic waste, and radioactive waste (Chartier et al., 2024; Janik-Karpinska et al., 2023).

General waste: This class of BMW is deemed safe due to the absence of any significant biological, chemical, radioactive, or physical hazards. Approximately 85% of the waste produced by healthcare activities belongs to this category, including food waste, paper, cardboard, wood materials, kitchen waste, and packaging materials (Sharma et al., 2018).

As earlier mentioned, the remaining 15% is classified as hazardous as they can cause significant threats to human health and the environment because of their characteristic microbial, physical, or chemical properties (Chartier et al., 2024). Hence, the following classes;

Infectious waste: These are considered infectious as they potentially transmit infections to HCWs, patients, and the public. Examples of waste that fall under this category are those contaminated with blood and bodily fluids, like discarded diagnostic samples. It also includes waste from laboratory work involving infectious agents, such as autopsies and experiments with infected animals. Furthermore, it encompasses waste from patients with infections, such as swabs, bandages, and disposable medical devices (Kanyal et al., 2021; Lambodaran et al., 2020). The main concern with this category of BWM is its ability to transmit infections. As such, autoclaving, incineration, and secure landfill disposal are critical components of effectively managing infectious waste (Chartier et al., 2014).

Pathological waste: comprising this category are human tissues, organs, fluids, body parts, postmortem examinations, the delivery of placenta or fetus, and contaminated animal carcasses. Ethical and cultural sensitivities, especially regarding the disposal of human body parts and fatal remains, must be taken into consideration by the management (Gupta, 2017; Subramanian et al., 2021). As such, incineration is frequently employed as a disposal method to ensure destruction (Gupta, 2017; Subramanian et al., 2021).

Sharps waste covers sharp instruments employed in patient care, medical research, or industrial laboratories. These instruments include hypodermic needles, syringes, disposable scalpels and blades, Pasteur pipettes, broken glassware, glass vials, and surgical knives (Jayakumar et al., 2022). The primary hazards include puncture wounds and the potential spread of blood-borne diseases. Consequently, the risk of injury and infection can be mitigated by regular training in the appropriate disposal of sharps to minimize accidents.

Chemical waste: These are substances deemed dangerous due to their toxic, corrosive, flammable, reactive, and genotoxic characteristics. Some examples of chemical BMWs include solvents and reagents used in laboratory preparations, disinfectants such as formaldehyde (which contributes significantly to chemical waste in hospitals), sterilants and heavy metals (HMs) found in medical devices (such as mercury in broken thermometers and batteries), as well as pharmaceutical waste like expired, unused, and contaminated drugs and vaccines, spent chemicals, and scrap amalgam (Chandrappa & Das, 2012; Kanyal et al., 2021).

The issue of chemical waste and its effects on the environment is becoming more worrisome, with particular attention being paid to water contamination, risks to human health, and the emergence of antibiotic-resistant bacteria. Hence, it is critical to implement proper labeling, storage, and disposal to prevent accidents and contamination of the environment (OSHA, 2012). Various approaches to handling chemical waste include manufacturer return programs, chemical neutralization, and high-temperature incineration (Chartier et al., 2014).

Cytotoxic/genotoxic waste: This includes genotoxic substances, which are hazardous compounds that can cause mutations, birth defects, or cancer. Examples include cytotoxic drugs used in cancer treatment and their by-products. As such, exposure to cytotoxic waste can lead to severe health hazards such as cancer and genetic mutations (Chartier et al., 2014). Therefore, specialized treatment methods such as chemical deactivation and incineration are necessary for safe management (Chartier et al., 2014) to ensure the safety of HCWs and the environment.

Radioactive waste: This pertains to waste that contains radioactive substances, which are commonly produced in hospitals, laboratories, and research facilities. Radionuclides in polluted diagnostic or therapeutic substances may be included (Sharma et al., 2018). It is imperative to follow strict regulatory standards to ensure the safe management of radioactive waste. This involves employing various measures, including shielding, containment, and disposal in specialized facilities, to prevent any risks of radiation exposure. The primary objective is to safeguard the well-being of HCWs, patients, and the environment.

Conclusively, it is worth noting that while only 15% of BMWs are classified as hazardous, it is still essential to recognize the potential risks associated with them, regardless of the quantity. This is due to their infectious, pathological, and cytotoxic properties. Hence, categorizing BMWs is critical in effectively handling and managing BMW waste.

8.4 Biomedical waste classification and implications for waste management practices

Categorizing BMWs significantly impacts management practices, improving efficiency and safety in healthcare settings. Proper classification is crucial in facilitating efficient waste separation at its origin, ensuring that infectious and hazardous substances are correctly identified and managed. Jaiswal et al. (2024) emphasize the significance of following the BMW management guide, which involves using color-coded bins to separate waste and prevent exposure risk and environmental pollution (Jaiswal et al., 2024). According to Joseph, integrating color-coded containers and improved classification systems greatly aids in segregating waste and mitigating infection risk and injury among HCWs, patients, and the community (Joseph, 2017).

Additionally, ensuring correct classification is of utmost importance, as any errors can result in a greater quantity of waste that requires specialized treatment, ultimately leading to higher expenses and adverse environmental impacts (Rajan et al., 2018). Similarly, understanding the different classes of BMWs and their inherent dangers is vital for implementing safety measures during handling, sorting, and disposal (Verma & Yadav, 2023). These measures include using personal protective equipment (PPE), regular training, and strict adherence to safety protocols (Al-Khatib et al., 2016; Verma & Yadav, 2023) to safeguard HCWs and waste handlers against potential exposure to hazardous materials.

8.5 Impact of COVID-19 on biomedical waste management

BMW management was significantly impacted on a global scale by the COVID-19 pandemic, leading to the worsening of existing challenges and the emergence of new ones. Thus, immediate adaptation is necessitated for developed and developing nations. Some of the notable impacts of COVID-19 on BMW management include the following:

8.5.1 Increased volume of waste

Evidence suggests the volume of BMW, which includes masks, gloves, aprons, face shields, and vaccine waste, experienced a substantial surge during the pandemic (Dehal et al., 2022; Kumar & Sharma, 2023; Olaniyi et al., 2021; Shefali & Kwon, 2023). In their study, Singh et al. (2022) emphasize the link between the expanded usage of these materials and the consequent surge in waste production, which poses significant challenges for waste management systems. For example, a study by Fadaei (2021) found that BMW increased by 350% to 500% in several countries, resulting in a strain on existing waste management systems (Kuntari et al., 2023; Kumar & Sharma, 2023). The rise caused a shift in the structure of BMW and exposed the deficiencies in handling such waste, creating significant environmental risks (Shefali & Kwon, 2023).

8.5.2 Handling and safety concerns

In various countries, particularly developing countries, HCWs frequently lack the training and awareness to manage BMWs effectively. This discrepancy became apparent during the pandemic, as the volume of waste exceeded the capabilities of current waste management systems. These difficulties have resulted in the improper management and disposal of infectious waste, which in turn raises the likelihood of infection transmission among HCWs, waste handlers, and the general population (Olaniyi et al., 2021), underscoring the need for regular training and the rigorous implementation of safety measures to safeguard public health (Khan et al., 2022).

8.5.3 Strain on waste treatment facilities

Similarly, waste treatment facilities faced enormous pressure due to the rise in waste generation. The sudden rise in waste production, along with limited funds and a lack of awareness among hospital administrators, has exposed major flaws in the current infrastructure (Vandana et al., 2020), indicating the inability of the existing waste management infrastructure to cope with the surge, thus requiring urgent upgrades and expansion (Kumar & Sharma, 2023). Also, a study conducted in Indonesia offers an illustrative example of how waste management was significantly influenced. The study found that the technical and financial aspects were particularly affected, resulting in increased expenses for waste transportation and a notable increase in CO_2 emissions caused by the larger quantity of waste. However, despite this, recycling activities had a minor impact on reducing emissions, as noted by Maziya et al. (2023).

8.5.4 Regulatory challenges

The enforcement of regulatory frameworks for BMW management is frequently lacking, even in everyday situations. The pandemic made these issues worse, as the amount of waste generated surpassed the capacity of existing regulatory and enforcement systems (Khan et al., 2022). For example, in Dakar, Senegal, the regulations governing BMW management, while having undergone improvements over time, faced challenges in keeping up with the rising demand, resulting in improper waste disposal and increased health risks (Dieng et al., 2022). Similarly, in India, the regulatory framework encountered comparable difficulties as the preexisting regulations, which were already inadequately enforced, were insufficient in effectively managing the substantial rise in BMW during the pandemic (Dehal et al., 2022). The importance of having regulatory frameworks that are adaptable and responsive to abrupt changes in waste generation patterns was highlighted by this situation.

The COVID-19 pandemic profoundly impacted BMW's management worldwide, exposing weaknesses and requiring immediate adjustments. The overwhelming increase in BMWs exposed the inadequacy of numerous waste management systems, emphasizing the critical need for robust infrastructure, well-designed regulations, and inventive technologies to guarantee the well-being of the public and the environment during similar emergencies.

8.6 Issues and challenges of biomedical waste management

As previously discussed, the outbreak of the COVID-19 pandemic has had a profound effect on waste management systems across the globe, with severe environmental and public health consequences, further complicating the already challenging task of BMW management. Thus, it underlines the importance of adopting efficient waste management practices to safeguard public health and the environment (Goswami et al., 2021). Among the significant changes facing BMW management are better discussed under the following;

8.6.1 Resource constraints

Many nations face challenges in managing BMW effectively because of limited financial resources, technology, and infrastructure. As a result, they struggle with inadequate waste segregation, handling, treatment, and disposal facilities (Bandaso & Ayuningtyas, 2019; Khan et al., 2019b; Nawaz & Subhani, 2023). Inadequate waste management facilities result in the improper handling of hazardous substances, which creates substantial health risks for HCWs and the public. For example, research on the management practices of BMW in developing nations reveals that these countries often face challenges in effectively segregating, collecting, handling, and disposing of waste, which poses significant health risks and causes environmental pollution (Nawaz & Subhani, 2023). Previous studies have demonstrated that inadequate facilities and proper disposal methods are significant challenges facing BMW management in several countries (Patan & Mathur, 2016).

8.6.2 Inadequate training and awareness

HCWs and waste handlers often lack proper training and awareness of safe BMW practices. The situation is worsened by the absence of continuous professional development initiatives and insufficient sharing of information regarding safety protocols (Baaki et al., 2022; Mitiku et al., 2022; Olaniyi et al., 2021). For example, a study in South Africa's Vhembe district revealed that HCWs faced significant hurdles in effectively managing BMW, including a poor understanding of waste management guidelines, insufficient training, and a lack of proper waste management equipment (Olaniyi et al., 2021). Similarly, inadequate handling and disposal practices for BMW in selected hospitals have been observed in Ahmedabad, India, as Deva et al. (2019) indicated. Therefore, highlighting the importance of continuous education and training in efficient BMW management for BMW handlers and HCWs will help reduce potential risks and improve management practices.

8.6.3 Poor regulatory frameworks

In countries, there are reports of a lack of adequate regulations or their poor enforcement when it comes to managing BMW, leading to inconsistent practices and the unauthorized recycling and disposal of dangerous materials, thus posing a significant risk to human health and the environment (Bandaso & Ayuningtyas, 2019). For example, studies indicate that even though various state-level regulations have been put in place. Numerous international treaties have been signed, but many healthcare facilities, especially in low- and middle-income countries, still lack adequate and efficient BMW management systems (Awodele et al., 2016; Dieng et al., 2020; Oruonye & Ahmed, 2020; Singh & Singh, 2018; Uchechukwu et al., 2017), emphasizing the significance of international cooperation in establishing universal guidelines and exchanging successful strategies to improve the global approach to managing BMW's challenges.

8.6.4 Technological deficiencies

The absence of advanced waste treatment methods like autoclaving, steam sterilization, and pyrolysis hampers the safe and efficient disposal of BMW waste. This is because the existing technologies, such as incinerators, are often poorly designed (Nemathaga et al., 2008) and outdated (Castellucci et al., 2022; Manga et al., 2011), leading to incomplete combustion and subsequent adverse environmental impacts. However, the prevailing belief is that the high cost of using and maintaining advanced technologies makes this more of an economic problem. In contrast, in most industrialized nations, there is a legal responsibility for institutions that generate BMW to handle their waste appropriately. Consequently, specific systems are in place to manage different types of waste, and the hazardous waste produced is continuously monitored (Kwikiriza et al., 2019).

8.6.5 Poor socioeconomic status

While developed countries have robust regulations and mechanisms to enforce them, developing nations face difficulties implementing and enforcing waste management policies due to various socioeconomic challenges (Bataduwaarachchi & Weeraratne, 2016). Insufficient allocation of funds for waste management infrastructure and personnel is a common issue faced by many healthcare facilities due to financial constraints, leading to informal waste disposal methods, which are both unsafe and harmful to the environment (Baaki et al., 2022). Also, a recent study on waste management challenges in developing countries found that limited financial resources worsen these issues, posing challenges in implementing efficient waste management systems (Bataduwaarachchi et al., 2018; Singh & Gupta, 2022).

Even developed nations encounter challenges despite sophisticated systems, particularly in the COVID-19 pandemic. The increase in BMW has significantly pressured existing waste management systems, prompting the need for rapid adaptations and innovations. Developing countries are grappling with issues such as inadequate infrastructure and limited regulatory enforcement, while developed countries face challenges associated with regulatory complexity and high costs. Developing solutions specifically designed to address the unique combination of local socioeconomic conditions, technological capabilities, and regulatory environments in both situations is crucial. Thus, it is vital to take a multifaceted approach, which involves implementing policies, advancing technology, and promoting awareness among stakeholders to tackle these challenges worldwide effectively.

8.7 Environmental and public health impact of biomedical waste

Globally, managing BMW poses significant challenges with wide-ranging implications for human health and the environment (Chudasama et al., 2013; Mane et al., 2016; Singh et al., 2014). The health risks of BMWs are extensive and affect humans and the environment. These risks include direct exposure to toxic chemicals and indirect injuries caused by the pollution of land, water, and air with waste materials from BMWs (Javid & Sharma, 2019). Therefore, it is increasingly vital to comprehend the negative impact of BMW on the environment and the potential consequences for public health.

8.8 Biomedical waste and public health

Generally, BMWs have specific qualities that make them one of the riskiest types of waste. If not managed properly, they can pose a severe threat to health, leading to outbreaks of infection and poisoning (Marceta & Nadj, 2018). Additionally, considering the widespread transmission of COVID-19 and the subsequent rise in BMW generation rates, BMW has emerged as a significant public health issue (Das et al., 2021), with the following concerns:

8.9 Infection risk

According to Makajic-Nikolic et al. (2016), BMW poses a health risk due to the potential presence of various infectious agents. These agents can be transmitted through different routes, including contact with damaged skin (rubbing, puncturing, or cutting), inhalation, mucous membranes, or ingestion (Chartier et al., 2014; Meshram & Mhatre, 2022; Nayak et al., 2012). This evidence suggests that BMWs may be capable of transmitting over 30 harmful blood-borne pathogens (Yazie et al., 2019), with potential risks of infection (Park et al., 2009; Udofia et al., 2017). According to Singh et al. (2020b), mishandling of BMWs, particularly by BMW managers, has been estimated to be responsible for approximately 30% of hepatitis B cases, 1%–3% of hepatitis C cases, and 0.3% of HIV cases. Furthermore, studies indicate that the prevalence of HBV and HCV is significantly higher among BMW handlers compared to non-BMW handlers or the general population (Akpieyi et al., 2014; Doddaiah et al., 2013; Janik-Karpinska et al., 2023). Variations in the frequency and duration of exposure may explain this and underscore the necessity of proper safety training and monitoring to enforce compliance with the appropriate standards.

8.10 Chemical intoxication

According to Chartier et al. (2014), BMWs typically contain certain levels of unwanted or expired chemicals and pharmaceutical by-products, and these substances pose a health risk, even in small quantities. Observation shows that chemical waste adversely affects human health (Doggalli, 2014). The primary consequence of contact with these wastes is intoxication due to exposure to corrosive, flammable, or reactive chemicals such as formaldehyde and other volatile substances (Doggalli, 2014), resulting in injuries to the eyes, skin, or respiratory tract membranes. Additionally, exposure to cytotoxic waste is considered hazardous, as direct contact can result in the development of a rash, dermatitis, irritation of the skin, ulceration of the mucous membranes, and irritation of the eyes or throat (Shahrasbi et al., 2014), while extended or repeated contact poses serious health consequences, including infertility, genital deformities, hormone-related cancers, genetic mutations, skin irritation, asthma, and neurological disorders (Capoor & Bhowmik, 2017; Javid & Sharma, 2019).

8.11 Sharps-related injuries

The management of BMW also poses severe concerns about injuries caused by sharps. According to a report, approximately 16 billion injections are administered worldwide yearly (Janik-Karpinska et al., 2023), with evidence suggesting unsafe needle disposal, which increases the risk of injuries (Anagaw et al., 2012), blood-borne infections (Meshram & Mhatre, 2022; Odumosu, 2015; Yazie et al., 2019), and opportunities for needle reuse (Odumosu, 2015). Likewise, BMW handlers face an existential risk of needle-stick injuries, including other biochemical hazards, while scavenging at disposal sites. This also applies to those handling and manually sorting BMW from healthcare facilities, especially in low- and middle-income nations. For instance, studies have demonstrated that individuals involved in handling BMWs face a heightened risk of contracting infections such as cholera, typhoid fever, hepatitis, tuberculosis, bronchitis, and skin infections due to unsafe practices (Cebe et al., 2013; Joseph et al., 2015; Meshram & Mhatre, 2022).

8.12 Risk of antimicrobial resistance

Improper disposal of antibiotics and other pharmaceutical waste contributes to the growth and transmission of antimicrobial resistance. The release of antibiotics from BMWs into the environment can lead to changes in microbial ecosystems and the proliferation of antibiotic resistance (Kanyal et al., 2021; Manzoor & Sharma, 2019; Shakira et al., 2018), posing a threat to public health (Odumosu, 2015). For example, evidence points to the introduction of multidrug resistant microbes into different water sources in the environment due to poor handling and disposal of BMW (Marathe et al., 2017; Odonkor & Addo, 2018), thus transferring resistance genes from environmental bacteria to human pathogens (Odonkor & Mahami, 2020). This poses a severe public health concern as it undermines the efficacy of antibiotics in treating infections. Likewise, studies have demonstrated the emergence of antibiotic resistance in certain bacterial strains, resulting in the accumulation of antibiotics in the environment (Chi et al., 2020; Forsberg et al., 2012; Shaaban et al., 2018), emphasizing the negative impact of pharmaceuticals on the environment. Consequently, it is essential to adequately decontaminate chemical and pharmaceutical waste before its disposal in landfills to prevent the interaction between environmental bacteria and antibiotics, resulting in antibiotic-resistant capabilities (Chi et al., 2020; Forsberg et al., 2012; Shaaban et al., 2018).

8.13 Environmental impact of biomedical waste

Over the years, BMWs have been implicated in environmental degradation, either directly or indirectly, due to improper handling of BMWs, thus hampering the sustainability of water, air, and soil quality (Babu et al., 2009; Javid & Sharma, 2019). Consequently, this mismanagement has led to environmental issues such as air pollution, water pollution, and soil pollution (Bhatia & Paul, 2017). However, the environmental consequences of improperly disposing of BMWs vary depending on the disposal method, hence the following environmental implications.

8.14 Air pollution risk

The incineration of BMW waste without proper control measures has been shown to generate dioxins, furans, and similar compounds, as highlighted in various studies (Gautam et al., 2010; Lekwot et al., 2012; Sharma et al., 2013a; Verma, 2014). This uncontrolled incineration leads to the contamination of the surrounding air, posing a severe threat to public health (Subramanian et al., 2021), with respiratory diseases among residents being a prominent concern (Ephraim et al., 2013; Vimercati et al., 2016). Additionally, reports suggest these toxicants have properties that can cause chronic illnesses, such as mutations, cancer, or birth defects (Gautam et al., 2010; Sharma et al., 2013a). This is particularly worrisome among developing countries due to the uncontrolled burning of BMWs, resulting in significant dioxin emissions (Ananth et al., 2010), emphasizing the necessity to adopt alternative measures for treating BMWs to mitigate environmental risks and ensure the well-being of the public.

8.15 Water pollution

Like other environmental media, improper disposal of BMWs has a detrimental impact on water quality due to contaminants like cadmium, lead, aluminum, vanadium, and iron. These pollutants can leach into the groundwater from BMW dump sites, particularly in unlined landfills where leachate can dissipate into the groundwater (Al Raisi et al., 2014). In this scenario, HMs and toxic hydrocarbon pollutants pose a significant risk to surface and groundwater pollution (Al Raisi et al., 2014; Heera & Rajor, 2014). For example, studies have shown that the wastewater from BMW contains harmful substances that are genotoxic and cytotoxic, posing a potential threat to both aquatic life and humans, mainly through the contamination of drinking water (Sharma et al., 2013b; Kori et al., 2020). Hence, the improper discharge of untreated or poorly treated wastewater from BMWs into natural water sources is a concern for the environment and public health and should be adequately addressed.

8.16 Soil pollution

Additionally, the disposal of BMWs without proper regulation has significant consequences for soil quality, as mixing various contaminants with the soil can alter its composition, reduce fertility, and cause an overall decline in quality (Garau et al., 2019). As a result, these pollutants can potentially infiltrate the food chain and groundwater. For instance, a study in southwestern Nigeria demonstrated that the concentration of HMs, such as zinc, lead, chromium, and cadmium, was higher in the soil samples and bottom ash of a hospital incinerator when compared to municipal dumpsites, indicating that the disposal of BMWs from healthcare facilities leads to significant soil pollution (Auta & Morenikeji, 2013). Research investigating environmental contamination near BMW incinerators has consistently found soil pollution with dioxins and HMs exceeding typical background concentrations, as indicated by multiple studies (Abidemi & Theresa, 2015; Ali et al., 2014; Amfo-Otu et al., 2015; Ansari et al., 2019; Javid & Sharma, 2019; Vincent et al., 2016). This pollution carries risks for agriculture, wildlife, and human health, highlighting the need for strict waste management practices to address the situation.

8.17 Climate change impact

The practices of managing BMWs, specifically incineration, have a detrimental impact on climate change, as they release greenhouse gases (GHGs), including carbon dioxide and methane, into the atmosphere, thus contributing to global warming. For example, Sharma et al. (2013a), in a review of the detrimental impacts of BMW incineration, documented emission of GHGs (e.g., sulfur dioxides, nitrogen oxides, and carbon tetrachloride), exacerbating greenhouse effect and contribute to the smog formation, global warming, acidification, and ozone destruction (Aymen & Firas, 2018; Blahuskova et al., 2019; Eckelman & Sherman, 2016). According to Health Care Without Harm, an international organization, the healthcare industry is the fifth-largest contributor to GHG emissions globally, accounting for approximately 4.4% of total net emissions (Karliner et al., 2019). A similar report revealed that the combustion of 1 kg of BMW produces approximately 3 kg of CO_2, making the incineration of BMWs a significant contributor to global warming due to the release of substantial amounts of GHGs (Wyssusek et al., 2018). More so, studies indicate that the improper disposal of a single intravitreal injection leads to the emission of 0.05 kg CO_2e (Power et al., 2021), underscoring the role of inadequate BMW handling in contributing to climate change and the need for sustainable and environmentally friendly waste management solutions to combat climate change.

8.18 Future directions

While BMW has faced mismanagement and improper disposal in various countries, it is crucial to recognize that there are possible solutions to address these challenges. The key lies in embracing advanced technologies for BMW treatment and adopting sustainable concepts, which have been effectively implemented in several developed and industrialized nations.

8.19 Enhanced waste segregation practices

The key to successful BMW management is effectively segregating waste at its source. This entails classifying the waste into different categories, such as infectious, noninfectious, chemical, or radioactive, to ensure proper handling. Compliance can be improved by implementing color-coded bins and giving HCWs clear guidelines. As such, to foster adherence to segregation protocols and mitigate the risk of contamination, it is crucial to implement ongoing training programs and regular monitoring (Bansod & Deshmukh, 2023). These measures contribute to reducing the volume of waste that necessitates specialized treatment, in alignment with WHO guidelines on standardizing waste segregation practices and integration into the national regulations for BMW management (Chartier et al., 2014).

8.20 Adoption of nonincineration technologies

Considering the substantial air pollution and the related environmental and public health hazards caused by incineration, it is imperative to prioritize alternative waste treatment methods. Technologies like autoclaving, microwave treatment, and plasma pyrolysis provide efficient waste sterilization without the negative environmental impact caused by incineration. This equipment offers several benefits, such as decreased environmental degradation, negligible health impacts, safe handling of treated wastes, eco-friendly operations, reduced microbial load, and improved disposal methods (Ali et al., 2017; Mmereki, Baldwin, Li, & Liu, 2017; Manzoor & Sharma, 2019), thereby mitigating the negative environmental and health consequences.

8.21 Strengthening regulatory frameworks and enforcement

Robust regulatory frameworks are also essential for successful BMW management, providing clear guidelines for healthcare facilities, waste handlers, and regulatory bodies to fulfill their roles effectively. Governments must reform and enforce BMW management rules, including stringent penalties for noncompliance. Additionally, there is a need to create standardized protocols for BMW treatment and disposal to ensure consistent practices across different regions (Manzoor & Sharma, 2019).

8.22 Promoting research and development

Investing in research and development cannot be overstated when finding innovative and sustainable solutions for effective BMW management. The research should prioritize the development of environmentally friendly waste treatment technologies, investigate the environmental consequences of different disposal methods, and explore waste-to-energy conversion possibilities. Partnerships among academic institutions, government agencies, and the private sector can foster progress in BMW management technologies (Vincent et al., 2016), including creating environmentally friendly biomedical products that do not necessitate sophisticated pollution control methods for their disposal.

8.23 Public awareness and education

Fostering community support and ensuring compliance with waste management practices requires raising awareness about BMW's environmental and health impacts. Public education initiatives can educate individuals on the significance of proper waste disposal and the dangers linked to BMW. In addition, healthcare facilities should actively collaborate with local communities to raise awareness about proper disposal methods and decrease environmental pollution (Kanyal et al., 2021). BMW management principles in school curricula have been advocated to cultivate early comprehension and ensure safe BMW practices (Bataduwaarachchi et al., 2018).

8.24 Conclusion

Mitigating the significant health and environmental impacts of BMW is crucial, especially considering the challenges of the post-COVID-19 pandemic. The rise in healthcare use and subsequent increase in BMW places additional stress on current waste management systems, highlighting the need for effective management. The absence of advanced waste management technologies, weak regulatory framework, and limited economic capacity, among other factors, exacerbate the scenario, especially in low-resource countries. As a result, the mismanagement of BMW has resulted in extensive water, air, and soil pollution, exacerbating public health crises, environmental degradation, and the impacts of climate change. The urgency of this situation highlights the need for improved management practices and policy regulation at BMW.

Consequently, it is vital to implement a multifaceted approach that focuses on improving waste management practices, strengthening regulatory enforcement, conducting research, and raising public awareness. This is necessary to effectively address the challenges of BMW management and its impact on the environment and public health. Thus, there is a need to implement policies that emphasize integrating efficient management practices into healthcare planning and delivery activities, aligning with the prevailing global perspective that effective management is an integral part of healthcare delivery.

References

Abidemi, O. O., & Theresa, O. C. (2015). Environmental fate of heavy metals in soil of Ido-Osun waste dump site, Osogbo, Osun, Nigeria. *American Journal of Environmental Protection, 3*(1), 1–4. doi:10.12691/env-3-1-1.

Akpieyi, A., Tudor, T. L., & Dutra, C. (2014). The utilisation of risk-based frameworks for managing healthcare waste: A case study of the National Health Service in London. *Safety Science, 72*, 127–132. doi:10.1016/j.ssci.2014.08.014.

Al Raisi, S. A. H., Sulaiman, H., Suliman, F. E., & Abdallah, O. (2014). Assessment of heavy metals in leachate of an unlined landfill in the Sultanate of Oman. *International Journal of Environmental Science and Development, 5*(1), 60–63.

Ali, M., Wang, W., Chaudhry, N., & Geng, Y. (2017). Hospital waste management in developing countries: A mini review. *Waste Management & Research: The Journal of the International Solid Wastes and Public Cleansing Association, ISWA, 35*(6), 581–592. doi:10.1177/0734242X17691344.

Ali, S. M., Pervaiz, A., Afzal, B., Hamid, N., & Yasmin, A. (2014). Open dumping of municipal solid waste and its hazardous impacts on soil and vegetation diversity at waste dumping sites of Islamabad city. *Journal of King Saud University - Science, 26*, 59–65. doi:10.1016/j.jksus.2013.08.003.

Al-Khatib, I. A., Eleyan, D., & Garfield, J. (2016). A system dynamics approach for hospital waste management in a city in a developing country: The case of Nablus. *Palestine. Environmental Monitoring and Assessment, 188*(9), 503. doi:10.1007/s10661-016-5487-9 2.

Amfo-Otu, R., Kyerewaa, S. G., Ofori, E. A., & Sadick, A. (2015). Comparative study of heavy metals in bottom ash from incinerators and open pit from healthcare facilities in Ghana. *Octa Journal of Environmental Research, 3*(1), 50–56.

Anagaw, B., Shiferaw, Y., Anagaw, B., Belyhun, Y., Erku, W., Biadgelegn, F., ... Mulu, A. (2012). Seroprevalence of hepatitis B and C viruses among medical waste handlers at Gondar town Health institutions, Northwest Ethiopia. *BMC research notes, 5*, 55. https://doi.org/10.1186/1756-0500-5-55.

Ananth, A. P., Prashanthini, V., & Visvanathan, C. (2010). Healthcare waste management in Asia. *Waste Management*, 154e161. doi:10.1016/j.wasman.2009.07.018.

Andeobu, L., Wibowo, S., & Grandhi, S. (2022). Medical waste from COVID-19 pandemic—A systematic review of management and environmental impacts in Australia. *International Journal of Environmental Research and Public Health, 19*(3), 1381. doi:10.3390/ijerph19031381.

Ansari, M., Mohammad, H. E., Mahdi, F., & Ehsan, A. (2019). Dynamic assessment of economic and environmental performance index and generation, composition, environmental and human health risks of hospital solid waste in developing countries; A state of the art of review. *Environment International, 132*, 105073. doi:10.1016/j.envint.2019.105073.

Auta, T., & Morenikeji, O. (2013). Heavy metal concentrations around a hospital incinerator and a municipal dumpsite in Ibadan City, South-West Nigeria. *Journal of Applied Sciences and Environmental Management, 17*, 419–424. doi:10.4314/jasem.v17i3.11.

Awodele, O., Adewoye, A. A., & Oparah, A. C. (2016). Assessment of medical waste management in seven hospitals in Lagos, Nigeria. *BMC Public Health [Electronic Resource], 16*, 1–11. doi:10.1186/s12889-016-2916-1.

Aymen, A.-U. L. S. A., & Firas, A. L. B. (2018). Environmental impacts of medical waste treatment and management by burning inside health facilities. *International Journal of Civil Engineering and Technology, 9*(5), 41–53.

Baaki, T. K., Baharum, M. R., Pitt, M., & Davies, S. (2022). Associated factors of medical waste management practices in developing countries: A review. *International Journal of Environment and Waste Management, 30*, 1. http://doi.org/10.1504/ijewm.2022.128106.

Babu, B., Ramesh, A. K., Parande, R., Rajalakshmi, P., Suriyakala, & Volga, M. (2009). Management of biomedical waste in India and other countries: A review. *Journal of International Environmental Application and Science, 4*(1), 65–78.

Bandaso, A., & Ayuningtyas, D. (2019). Management of medical waste in developing countries: A systematic review. *Strengthening Hospital Competitiveness to Improve Patient Satisfaction and Better Health Outcomes*. doi:10.26911/the6thicph-FP.01.02.

Bansod, H., & Deshmukh, P. (2023). Biomedical waste management and its importance: A systematic review. *Cureus, 15*. doi:10.7759/cureus.34589.

Bataduwaarachchi, V., Thevarajah, R., & Weeraratne, C. (2018). Medication waste disposal practices among patients attending selected outpatient departments in a tertiary care institution: A cross-sectional survey. *International Journal of Basic & Clinical Pharmacology, 7*, 888–894.

Bataduwaarachchi, V., & Weeraratne, C. (2016). Global medication waste management practices: Challenges and opportunities in developing countries. *International Journal of Basic and Clinical Pharmacology, 5*, 2290–2294. doi:10.18203/2319-2003.IJBCP20164081.

Bendjoudi, Z., Taleb, F., Abdelmalek, F., & Addou, A. (2009). Healthcare waste management in Algeria and the Mostaganem department. *Waste Management, 29*(4), 1383–1387. doi:10.1016/j.wasman.2008.10.008.

Bhatia, D., & Paul, D. (2017). Biomedical waste management (BMW) and its awareness in India (in accordance with BMW Rules, 2016). In *ENVIS RP Centre* (p. 27). North Eastern Hill University.

Blahuskova, V., Vlcek, J., & Jancar, D. (2019). Study connective capabilities of solid residues from the waste incineration. *Journal of Environmental Management, 231*, 10481055. doi:10.1016/j.jenvman.2018.10.112.

Capoor, M. R., & Bhowmik, K. T. (2017). Cytotoxic drug dispersal, cytotoxic safety, and cytotoxic waste management: Practices and proposed india-specific guidelines. *Indian Journal of Medical and Paediatric Oncology (Official), 38*, 190–197. doi:10.4103/ijmpo.ijmpo_174_16.

Castellucci, A., Ragazzi, M., Berghi, F., Guirreri, D., & Rama, I. (2022). Biomedical solid waste management in low-income contexts: The case of Beira, Mozambique. *Ecology and the Environment*. doi:10.2495/wmei220091.

Cebe, A., Dursun, S., & Mankolli, H. (2013). Hospital solid wastes and their effect on the environment. *Journal of International Environmental Application and Science, 8*(5), 733–737.

Chandrappa, R., & Das, D. B. (2012). Chapter 6: Biomedical waste. In *Solid waste management* (pp. 147–175). Springer. doi:10.1007/978-3-642-28681-0_6.

Chartier, Y., Emmanuel, J., Pieper, U., Pruess, A., Rushbreak, P., Stringer, R., Townsend, W., Wilbum, S., & Zghondi, R. (2024). *Safe management of wastes from health-care activities* (2nd Ed). World Health Organization https://www.ircwash.org/sites/default/files/9789241548564_eng.pdf.

Chartier, Y., Emmanuel, J., Pieper, U., Prüss, A., Rushbrook, P., Stringer, R., . . . Wilburn (2014). In R. Zghondi (Ed.), *Safe management of wastes from health-care activities* (2nd ed.). World Health Organization https://apps.who.int/iris/handle/10665/85349.

Chi, T., Zhang, A., Zhang, X., Li, A.-D., Zhang, H., & Zhao, Z. (2020). Characteristics of the antibiotic resistance genes in the soil of medical waste disposal sites. *Science of the Total Environment, 730*, 139042. doi:10.1016/j.scitotenv.2020.139042.

Chudasama, R. K., Rangoonwala, M., Sheth, A., Misra, S. K. C., Kadri, A. M., & Patel, U. V. (2013). Biomedical Waste Management: A study of knowledge, attitude and practice among health care personnel at tertiary care hospital in Rajkot. *Journal of Research in Medical and Dental Science, 1*, 17–22.

Das, A. K., Islam, M. N., Billah, M. M., & Sarker, A. (2021). COVID-19 pandemic and healthcare solid waste management strategy - A mini-review. *The Science of the Total Environment, 778*, 146220. doi:10.1016/j.scitotenv.2021.146220.

Dehal, A., Vaidya, A. N., & Kumar, A. R. (2022). Biomedical waste generation and management during COVID-19 pandemic in India: Challenges and possible management strategies. *Environmental Science and Pollution Research, 29*, 14830–14845. doi:10.1007/S11356-021-16736-8.

Dehghani, M. H., Ahrami, H. D., Nabizadeh, R., Heidarinejad, Z., & Zarei, A. (2019). Medical waste generation and management in medical clinics in South of Iran. *MethodsX, 6*, 727–733. doi:10.1016/j.mex.2019.03.029.

Deva, L., Shah, C., Yagnik, B., & Solanki, H. (2019). Principles and practices of biomedical waste management: A case study of selected hospitals of Ahmedabad city. *Universal Journal of Environmental Research and Technology. Universal Review, VIII*(IV), 441–459.

Dieng, C., Mberu, B., Dimbuene, Z. T., & Faye, C. (2022). Biomedical waste management in Dakar, Senegal: Legal framework, health and environment issues; policy and program options. *Cities & Health*. https://www.academia.edu/download/98457670/23748834.2020.178622820230209-1-1jzlz2a.pdf.

Dieng, C., Mberu, B., Dimbuene, Z. T., Faye, C., Amugsi, D., & Aboderin, I. (2020). Biomedical waste management in Dakar, Senegal: Legal framework, health and environment issues; policy and program options. *Cities & Health, 6*(1), 208–222. doi:10.1080/23748834.2020.1786228.

Doddaiah, V., Janakiram, K., & Javagal, S. (2013). Seroprevalence of Hepatitis B Virus and Hepatitis C Virus in healthcare workers-aims, Bg Nagara. *American Journal of Life Sciences, 1*, 145–149.

Doggalli, D. N. (2014). Hazards and public health impacts of hospital waste. *Journal of Applied Research, 4*, 384. doi:10.15373/2249555X/June2014/120.

Eckelman, M. J., & Sherman, J. (2016). Environmental impacts of the U.S. health care system and effects on public health. *PLoS One, 11*(6), e0157014. doi:10.1371/journal.pone.0157014.

Ephraim, I. P., Ita, A., & Eusebius, O. (2013). Investigation of soils affected by burnt hospital wastes in Nigeria using PIXE. *Springer Plus, 2*(1), 208. doi:10.1186/2193-1801.

Fadaei, A. (2021). Study of solid waste (municipal and medical) management during the COVID-19 pandemic: A review study. *Reviews on Environmental Health, 37*(4), 559–566. doi:10.1515/reveh-2021-0092.

Forsberg, K. J., Reyes, A., Wang, B., Selleck, E. M., Sommer, M. O. A., & Dantas, G. (2012). The shared antibiotic resistance of soil bacteria and human pathogens. *Science, 337*, 1107. doi:10.1126/science.1220761.

Garau, G., Porceddu, A., Sanna, M., Silvetti, M., & Castaldi, P. (2019). Municipal solid wastes as a resource for environmental recovery: Impact of water treatment residuals and compost on the microbial and biochemical features of As and trace metal-polluted soils. *Ecotoxicology and Environmental Safety, 174*, 445–454. doi:10.1016/j.ecoenv.2019.03.007.

Gautam, V., Thapar, R., & Sharma, M. (2010). Biomedical waste management: Incineration vs. environmental safety. *Indian Journal of Medical Microbiology, 28*(3), 191–192. doi:10.4103/0255-0857.66465.

Goswami, M., Goswami, P. J., Nautiyal, S., & Prakash, S. (2021). Challenges and actions to the environmental management of bio-medical waste during COVID-19 pandemic in India. *Heliyon, 7*(7), e07310. doi:10.1016/j.heliyon.2021.e07310.

Gupta, S. (2017). Biomedical waste management in India- A review. *Journal of Advanced Medical and Dental Sciences Research, 5*(4), 99–102. doi:10.21276/jamdsr.2017.5.4.22.

Heera, S., & Rajor, A. (2014). Bacterial treatment and metal characterization of biomedical waste ash. *Journal of Waste Management*, 1–7. doi:10.1155/2014/956316.

Insa, E., Zamorano, M., & Lopez, R. (2010). Critical review of medical waste legislation in Spain. *Resources, Conservation and Recycling, 54*, 1048e1059. doi:10.1016/j.resconrec.2010.06.005.

Jaiswal, M. S. V., Bhansali, M. K. P., Ugle, M. K. S., Khan, M. N., & Reddy, K. R. (2024). Survey of biomedical waste management. *International Journal of Advanced Research in Science, Communication and Technology, 4*(2). doi:10.48175/ijarsct-15788.

Janik-Karpinska, E., Brancaleoni, R., Niemcewicz, M., Wojtas, W., Foco, M., Podogrocki, M., & Bijak, M. (2023). Healthcare waste: A serious problem for global health. *Healthcare (Basel, Switzerland), 11*(2), 242. doi:10.3390/healthcare11020242.

Javid, M., & Sharma, M. (2019). Impact of biomedical waste on environment and human health. *Environmental Claims Journal*. doi:10.1080/10406026.2019.1619265.

Jayakumar, A., Veerakumar, V., Pradeepsankar, S., Santhosh, K., Mohan, A. M., & Mahadevan, S. (2022). Biomedical waste management in prosthodontic dentistry. *International Journal of Health Sciences, 6*(S2), 1727–1737. doi:10.53730/ijhs.v6nS2.5303.

Joseph, J. (2017). Awareness and practice of biomedical waste management in Indian setting (2012-2016): Systematic review. *Research and Reviews: Journal of Medical and Health Sciences, 6*, 27–33.

Joseph, L., Paul, H., Premkumar, J. R., Paul, R., & Michael, J. (2015). Biomedical waste management: Study on the awareness and practice among healthcare workers in a tertiary teaching hospital. *Indian Journal of Medical Microbiology, 33*, 129–131. doi:10.4103/0255-0857.148411.

Kanyal, D., Butola, L., & Ambad, R. (2021). Biomedical waste management in India-A review. *Indian Journal of Forensic Medicine & Toxicology*. doi:10.37506/IJFMT.V15I2.14285.

Karliner, J., Slotterback, S., Boyd, R., Ashby, B., & Steele, K. (2019). Health care's climate footprint: How the health sector contributes to the global climate crisis and opportunities for action. *Health Care Without Harm*. https://noharm-global.org/sites/default/files/documents-files/5961/HealthCaresClimateFootprint_092319.pdf.

Khan, B. A., Khan, A. A., Ahmed, H., Shaikh, S. S., Peng, Z., & Cheng, L. (2019a). A study on small clinics waste management practice, rules, staff knowledge, and motivating factor in a rapidly urbanizing area. *International Journal of Environmental Research and Public Health, 16*(20), 4044.

Khan, B. A., Cheng, L., Khan, A., & Ahmed, H. (2019b). Healthcare waste management in Asian developing countries: A mini review. *Waste Management & Research, 37*, 863–875. doi:10.1177/0734242X19857470.

Khan, W. S., Asmatulu, E., Uddin, M. N., & Asmatulu, R. (2022). *Biological waste disposal. Recycling and reusing of engineering materials* (pp. 295–309). Elsevier. doi:10.1016/b978-0-12-822461-8.00004-8.

Kori, R., Saxena, A., & Mokhle, S. (2020). Bioassay Study of biomedical liquid waste (treated effluent) of Bhopal City, Madhya Pradesh, India. *International Journal of Research, 7*, 164–173.

Kumar, V., & Sharma, P. (2023). Processes for the treatment of biomedical wastes: Challenges and issues. In *Biomedical waste management* (pp. 119–137). Elsevier. doi:10.1016/b978-0-323-90909-9.00007-1.

Kuntari, N. A. N., Ulfa, M., & Jamal, A. (2023). The impact of COVID-19 pandemic on hospital waste management: A bibliometric analysis. *Avicenna Journal of Environmental Health Engineering, 10*(4), 1–12. doi:10.34172/ajehe.5319.

Kwikiriza, S., Stewart, A. G., Mutahunga, B., Dobson, A. E., & Wilkinson, E. (2019). A whole systems approach to hospital waste management in Rural Uganda. *Frontiers in Public Health, 7*, 136. doi:10.3389/fpubh.2019.00136.

Lambodaran, G., Ganesh, S., Shyam, S., & Prabhakar, M. (2020). Biomedical waste: A review. *Medico-Legal Update, 20*(4). doi:10.37506/MLU.V20I4.2171.

Lekwot, V., Nunyi, B., Ifeanyi, E., Okafor, C. I., & Adamu, B. (2012). Public health implication of improper hospital waste disposal in Zonkwa district of Zangon-kataf local government area, Kaduna state. *Journal of Research in Environmental Science and Toxicology, 1*, 6.

Makajic-Nikolic, D., Petrovic, N., Belic, A., Rokvic, M., Radakovic, J. A., & Tubic, V. (2016). The fault tree analysis of infectious medical waste management. *Journal of Cleaner Production, 113*, 365–373. doi:10.1016/j.jclepro.2015.11.022.

Mane, V., Nimbannavar, S. M., & Yuvaraj, B. Y. (2016). Knowledge, attitude, and practices on biomedical waste and its management among health care workers at a tertiary care hospital in Koppal, Karnataka, India. *International Journal of Community Medicine and Public Health, 3*, 2953–2957. doi:10.18203/2394-6040.ijcmph20163390.

Manga, V. E., Forton, O. T., Mofor, L. A., & Woodard, R. (2011). Health care waste management in Cameroon: A case study from the Southwestern region. *Resources, Conservation and Recycling, 57*, 108–116. doi:10.1016/j.resconrec.2011.10.002.

Manzoor, J., & Sharma, M. (2019). Impact of biomedical waste on environment and human health. *Environmental Claims Journal, 31*, 311–334. doi:10.1080/10406026.

Marathe, N. P., Pal, C., Gaikwad, S. S., Jonsson, V., Kristiansson, E., & Larsson, D. G. J. (2017). Untreated urban waste contaminates Indian river sediments with resistance genes to last resort antibiotics. *Water research, 124*, 388–397. https://doi.org/10.1016/j.watres.2017.07.060.

Marceta, M., & Nadj, I. (2018). Effect of medical waste on health of population and environment in the Republic of Serbia. *Research Reviews of the Department of Geography, Tourism and Hotel Management, 47*, 18. doi:10.5937/ZbDght1802094M.

Maziya, F. B., Luthi, R. S., Iresha, F. M., Abidin, A. U., & Kasam (2023). The influence of pandemic COVID-19 on hazardous waste management from hospital A in Yogyakarta. *IOP Conference Series: Earth and Environmental Science, 1257*(1), 1–10. doi:10.1088/1755-1315/1257/1/012014.

Meshram, L. N., & Mhatre, K. J. (2022). Impact of biomedical waste on environment and human health. *International Journal of Zoological Investigations, 8*(2), 484–489. doi:10.33745/ijzi.2022.v08i02.060.

Mitiku, G., Admasie, A., Birara, A., & Yalew, W. (2022). Biomedical waste management practices and associated factors among health care workers in the era of the covid-19 pandemic at metropolitan city private hospitals, Amhara region, Ethiopia, 2020. *PLoS One, 17*. doi:10.1371/journal.pone.0266037.

Mmereki, D., Baldwin, A. N., Li, B., & Liu, M. (2017). Healthcare waste management in Botswana: storage, collection, treatment and disposal system. *Journal of Material Cycles and Waste Management, 19*, 351–365.

Nawaz, N., & Subhani, M. (2023). Hazardous biomedical waste management scenario in developing countries. In *Handbook of environmental materials management* (pp. 377–389). Elsevier. doi:10.1016/b978-0-323-90463-6.00015-4.

Nayak, A., Chopra, R., Kaur, R., Eye, S., & Nayak, A. (2012). Biomedical waste management: A review. *Journal of Oral Health and Community Dentistry, 6*, 141–144. doi:10.5005/JOHCD-6-3-141.

Nemathaga, F., Maringa, S., & Chimuka, L. (2008). Hospital solid waste management practices in Limpopo Province, South Africa: A case study of two hospitals. *Waste Management, 28*(7), 1236–1245.

Odonkor, S. T., & Mahami, T. (2020). Healthcare waste management in Ghanaian hospitals: Associated public health and environmental challenges. *Waste Management & Research*, 1–9. doi:10.1177/0734242X2091.

Odonkor, S. T, & Addo, K. K (2018). Resistant Escherichia coli isolated from drinking water sources. *International Journal of Microbiology*, 7204013.

Odumosu, B. T. (2015). Biomedical waste: Its effects and safe disposal. Chapter 4. In R Chandra (Ed.), *Environmental waste management'. Boca Raton* (pp. 81–93). CRC Press. doi:10.1201/b19243-5.

Ojha, P. C., Satpathy, S. S., Ojha, A. K., Sukla, L. B., & Pradhan, D. (2022). Overcoming challenges due to enhanced biomedical waste generation during COVID-19 pandemic. *Science of the Total Environment, 832*, 155072. doi:10.1016/j.scitotenv.2022.155072.

Olaniyi, F. C., Ogola, J., & Tshitangano, T. (2021). Challenges of effective management of medical waste in low-resource settings: Perception of healthcare workers in Vhembe district healthcare facilities, South Africa. *Transactions of the Royal Society of South Africa, 76*, 81–88. doi:10.1080/0035919X.2021.1900949.

Oruonye, E., & Ahmed, A. (2020). Covid-19 and challenges of management of infectious medical waste in Nigeria: A case of Taraba State. *International Journal of Waste Resources, 10*, 1–5.

OSHA. (2012). Hazardous waste operations and emergency response. *Occupational Safety and Health Administration*. https://www.osha.gov/lawsregs/regulaôons/standardnumber/1910/1910.120.

Pachauri, A., Shah, P., Almroth, B. C., Sevilla, N. P., & Narasimhan, M. (2019). Safe and sustainable waste management of selfcare products. *BMJ, 365*. doi:10.1136/bmj.l1298.

Parida, V. K., Sikarwar, D., Majumder, A., & Gupta, A. K. (2022). An assessment of hospital wastewater and biomedical waste generation, existing legislations, risk assessment, treatment processes, and scenario during COVID-19. *Journal of Environmental Management, 308*, 114609. doi:10.1016/j.jenvman.2022.114609.

Park, H., Lee, K., Kim, M., Lee, J., Seong, S.-Y., & Ko, G. (2009). Detection and hazard assessment of pathogenic microorganisms in medical wastes. *Journal of Environmental Science and Health, Part A: Toxic/Hazardous Substances & Environmental Engineering, 44*, 995–1003. doi:10.1080/10934520902996898.

Patan, S., & Mathur, P. (2016). Assessment of biomedical waste management in government hospital of Ajmer city-a study. *Journal of Research in Pharmacy & Life Sciences, 3*(4), 316–324.

Power, B., Brady, R., & Connell, P. (2021). Analysing the carbon footprint of an intravitreal injection. *Journal of Ophthalmic and Vision Research, 16*, 367–376.

Rahman, M. M., Bodrud-Doza, M., Griffiths, M. D., & Mamun, M. A. (2020). Biomedical waste amid COVID19: Perspectives from Bangladesh. *The Lancet Global Health, 8*, e1262. doi:10.1016/S2214109X(20)30349-1.

Rajan, R., Robin, D., & Vandanarani, M. (2018). Biomedical waste management in Ayurveda hospitals – current practices and future prospectives. *Journal of Ayurveda and Integrative Medicine, 10*, 214–221. doi:10.1016/j.jaim.2017.07.011.

Shaaban, H., Alghamdi, H., Alhamed, N., Alziadi, A., & Mostafa, A. (2018). Environmental contamination by pharmaceutical waste: Assessing patterns of disposing unwanted medications and investigating the factors influencing personal disposal choices. *Journal of Pharmacology and Pharmaceutical Research, 1*, 003.

Shahrasbi, A. A., Afshar, M., Shokraneh, F., Monji, F., Noroozi, M., Ebrahimi-Khojin, M., ... Rajabi, M. (2014). Risks to health professionals from hazardous drugs in Iran: a pilot study of understanding of healthcare team to occupational exposure to cytotoxics. *EXCLI journal, 13*, 491–501.

Shakira, M., Hamna, K., Zainab, K., Saba, N., Sana, Z., & Noshabah, T. (2018). Hospital waste management: Execution in Pakistan and environmental concerns A review. *Environmental Contamination Reviews, 1*(1), 18–23. doi:10.26480/ecr.01.2018.18.23.

Sharma, P., Kumar, M., Mathur, N., Singh, A., Bhatnagar, P., & Sogani, M. (2013b). Health care industries: Potential generators of genotoxic waste. *Environmental Science and Pollution Research, 20*(8), 5160–5167. doi:10.1007/s11356-013-1500-9.

Sharma, P., Sharma, A., Nd, J., & Somani, P. (2018). A review on biomedical waste and its management. *Significances of Bioengineering & Biosciences*. doi:10.31031/SBB.2018.01.000522.

Sharma, R., Sharma, M., Sharma, R., & Sharma, V. (2013a). The impact of incinerators on human health and environment. *Reviews on Environmental Health, 28*(1), 67–72. doi:10.1515/reveh-2012-0035.

Sharma, H.B., Vanapalli, K.R., Cheela, V.S., Ranjan, V.P., Jaglan, A.K., Dubey, B. Goel, S., & Bhattacharya, J. (2020). Challenges, opportunities, and innovations for effective solid waste management during and post COVID-19 pandemic. Resources, Conservation and Recycling. 162, 105052. http://doi.org/10.1016/j.resconrec.2020.105052.

Shefali, & Kwon, J.-H. (2023). Generation, types, and impacts of biomedical waste during COVID-19: Indian context. *Environmental Analysis, Health and Toxicology, 38*(4), 1–13. doi:10.5620/eaht.2023019.

Singh, D. K., & Singh, A. K. (2018). An overview of the new legal regime of bio medical waste management in India. *Asian Journal of Multidimensional Research (AJMR), 7*, 32–45.

Singh, A., Kumari, R., Wakhlu, A., Srivastava, K., Wakhlu, A., & Kumar, S. (2014). Assessment of Bio-Medical Waste Management in a Government Healthcare Setting of North India. *International Journal of Health Sciences and Research, 4*, 203–208.

Singh, D., Aryan, Y., Chavan, D., & Tembhare, M. (2022). Mask consumption and biomedical waste generation rate during Covid-19 pandemic: A case study of central India. *Environmental Research, 210*, 112951. doi:10.1016/j.envres.2022.112951.

Singh, J., & Gupta, S. (2022). Waste management challenges in developing countries. In *Sustainable Waste Management: Policies and Case Studies* (pp. 29–42). CRC Press. doi:10.1201/9781003132349-2.

Singh, N., Tang, Y., & Ogunseitan, O. A. (2020b). Environmentally sustainable management of used personal protective equipment. *Environmental Science & Technology, 54*, 8500–8502. doi:10.1021/acs.est.0c03022.

Singh, N., Tang, Y., Zhang, Z., & Zheng, C. (2020a). COVID-19 waste management: Effective and successful measures in Wuhan, China. *Resources, Conservation, and Recycling, 163*, 105071. doi:10.1016/j.resconrec.2020.105071.

Subramanian, A., Thayalan, D., Edwards, A., Al-Malki, A. E., & Venugopal, A. (2021). Biomedical waste management in dental practice and its significant environmental impact: A perspective. *Environmental Technology and Innovation, 24*, 101807. doi:10.1016/j.eti.2021.101807.

Uchechukwu, E. E, IO, B., & CN, A. (2017). Investigating knowledge, attitude, and health care waste management by health workers in a Nigerian tertiary health institution. *Global Journal of Health Science, 9*, 222–232.

Udofia, E. A., Gulis, G., & Fobil, J. (2017). Solid medical waste: A cross-sectional study of household disposal practices and reported harm in Southern Ghana. *BMC Public Health [Electronic Resource], 17*, 464. doi:10.1186/s12889-017-4366-9.

Vandana, Rao, V. V., & Ghosh, S. K. (2020). Sustainable biomedical waste management—Case study in India. In *Urban mining and sustainable waste management* (pp. 309–320). Springer. doi:10.1007/978-981-15-0532-4_30.

Verma, A., & Yadav, S. K. (2023). Biomedical waste management. In S. K. Yadav, R. Gupta, & S. Singh (Eds.), *Clinical laboratory management*. Springer. doi:10.1007/978-3-031-46420-1_36.

Verma, R. (2014). Medical waste disposal: Incineration and non-incineration technology their effects and prospects. *Nature Environment*, 195e198.

Vimercati, L., Baldassarre, A., Gatti, M. F., De Maria, L., Caputi, A., Dirodi, A. A., … Bellino, R. M. (2016). Respiratory Health in Waste Collection and Disposal Workers. *International journal of environmental research and public health, 13*(7), 631. https://doi.org/10.3390/ijerph13070631.

Vincent, C., Singh, A. A. P., Supate, A., & Desai, N. (2016). Comparative study of heavy metal contamination at common biomedical waste treatment and disposal sites (incineration and deep burial) in Mumbai, Maharashtra, India. *International Journal of Health Sciences and Research, 6*(5), 415–419.

Windfeld, E. S., & Brooks, M. S.-L. (2015). Medical waste management – A review. *Journal of Environmental Management, 163*, 98–108. doi:10.1016/j.jenvman.2015.08.013.

Wise, J. (2022). COVID-19: Pandemic waste threatens human and environmental health, says WHO. *BMJ, 376*, o266. doi:10.1136/bmj.o266.

Wyssusek, K. H., Keys, M. T., & van Zundert, A. A. J. (2018). Operating room greening initiatives—The old, the new, and the way forward: A narrative review. *Waste Management & Research: The Journal for a Sustainable Circular Economy), 37*, 3–19. doi:10.1177/0734242X18793937.

Yazie, T. D., Tebeje, M. G., & Chufa, K. A. (2019). Healthcare waste management current status and potential challenges in Ethiopia: A systematic review. *BMC Research Notes, 12*, 285. doi:10.1186/s13104-019-4316-y.

You, S., Sonne, C., & Ok, Y. S. (2020). COVID-19's unsustainable waste management. *Science, 368*, 1438. doi:10.1126/science.abc7778.

Yu, H., Sun, X., Solvang, W. D., & Zhao, X. (2020). Reverse logistics network design for effective management of medical waste in epidemic outbreaks: Insights from the coronavirus disease (COVID-19) outbreak in Wuhan (China). *International Journal of Environmental Research and Public Health), 17*(5), 1770. doi:10.3390/ijerph17051770.

CHAPTER

9 Future frontiers in biomedical waste management: Emerging technologies and sustainable strategies for a postpandemic world

Suprava Sahoo, Jyotirmayee Lenka and Basudeba Kar

Centre for Biotechnology, Siksha O Anusandhan (Deemed to be University), Bhubaneswar, Odisha, India

9.1 Introduction

The emergence of SARS-CoV-2 in Wuhan, China, in December 2019 marked the beginning of a global health crisis that would transform virtually every aspect of human society. As the virus spread worldwide, nations implemented sweeping protective measures, including mandatory face mask policies, enhanced hygiene protocols, social distancing requirements, and widespread lockdowns—all aimed at controlling transmission rates and reducing healthcare system burden (Pokhrel & Chhetri, 2021).

This unprecedented public health emergency, widely regarded as the defining crisis of the 21st century, catalyzed far-reaching disruptions across economic, social, and environmental domains. The pandemic's impact extended beyond immediate health concerns, significantly hampering progress toward the 2030 Sustainable Development Goals (SDGs) and destabilizing global markets (Ranjbari et al., 2023). While the World Health Organization's recent declaration marking the end of COVID-19's pandemic status signals a return to normalcy, the crisis's lasting effects continue to reverberate through society. Though some environmental benefits emerged during lockdown periods, such as improved air quality and water clarity (Gardiner, 2020), the pandemic's impact on waste management systems reveals a more complex and troubling picture.

Waste management, intrinsically linked to environmental sustainability, economic efficiency, and public health, faces unprecedented challenges in the postpandemic landscape (Rodic & Wilson, 2017). The crisis has particularly strained municipal solid waste (MSW) management systems, creating new obstacles for sanitation workers and policy architects alike (Mallapur, 2020). Developing nations have been especially vulnerable, as suboptimal waste handling practices increased viral transmission risks

Biomedical Waste Management in the COVID-19 Era: A Global Scenario.
DOI: https://doi.org/10.1016/B978-0-443-21841-5.00003-5

through contaminated MSW. The United Nations Environment Program emphasizes that comprehensive waste management—encompassing collection, segregation, storage, transport, treatment, and disposal—has become crucial to global pandemic response strategies (UNEP, 2020).

Furthermore, the marine environment has suffered significant collateral damage, with plastic pollution reaching new heights during the pandemic. These materials, laden with chemical compounds including surfactants, dyes, and heavy metals, contribute to ocean eutrophication and pose severe threats to marine ecosystems. The accumulation of plastics in aquatic environments has created novel microbial habitats, termed "plastispheres," with far-reaching implications for marine ecological systems (Lenka et al., 2024).

Healthcare facilities' increased reliance on medical technology, combined with enhanced safety protocols, has generated unprecedented volumes of biomedical waste (BMW). The composition of this waste stream has shifted dramatically, now incorporating vast quantities of personal protective equipment (PPE) alongside traditional medical waste such as syringes, testing materials, and medical containers. The surge in plastic consumption, already trending upward due to population growth and industrialization, has accelerated sharply during the pandemic. Many essential protective items, including masks, gloves, and other PPE, utilize nonbiodegradable polymers like polyethylene (PE) and polypropylene (PP), contributing to long-term environmental accumulation (Barnes et al., 2009). Current projections suggest global waste generation could reach 3.4 billion tons annually, highlighting the urgency of addressing this growing crisis (Kaza et al., 2018).

Microplastics (MPs) represent a particularly insidious threat, serving as vectors for various contaminants, including heavy metals, flame retardants, and pharmaceutical residues. Their presence in ecosystems poses significant risks to both environmental and human health. While plastics offer undeniable benefits in terms of durability, cost-effectiveness, and versatility, their widespread use has contributed to the transformation of our biosphere into what researchers term a "terrestrial plastisphere." Addressing these challenges requires innovative approaches to waste management and a fundamental rethinking of our relationship with plastic materials (Priyadarshini et al., 2024). Despite the unavoidable increase in BMW, solid waste, and MPs during the pandemic, it is essential to draw lessons from this experience to better prepare for and prevent similar outbreaks in the future. Overall, this chapter aims to explore the future implications of waste management strategies in addressing the environmental impacts caused by the large volumes of waste generated during the pandemic.

9.2 Current waste generation scenario and urgency for management

Since the emergence of the COVID-19 pandemic, the demand for PPE has surged dramatically, leading to a significant increase in waste generation. This has been particularly evident with the proliferation of single-use items such as hand sanitizers, masks, gloves, and syringes, contributing to a notable rise in plastic waste (Sharma et al., 2020). Compounding this issue is the growing concern about the cleanliness of recyclable materials, which has further reduced the use of single-use plastics (Klemeš et al., 2020).

The PPE market is grappling with high risks of viral transmission, which makes the reuse of masks inadvisable. The universal mandate for face masks to mitigate airborne transmission has resulted in an increased waste burden (Bauchner et al., 2020). Many restaurants and cafes have opted for single-use packaging, limiting the use of reusable containers that could otherwise help minimize virus transmission (Naughton, 2020). Waste treatment systems designed for pre-COVID-19 waste are now ill-equipped to handle the drastic increases in waste generation. Evaluating the capacity of existing waste management systems during the pandemic is a critical area for scientific inquiry (Klemeš et al., 2020).

Globally, about 3 billion people lacked access to managed waste facilities, while approximately 2 billion lacked access to waste collection before the pandemic (Wilson et al., 2015). China, the largest manufacturer of PPE, is estimated to produce around 20 million masks daily, contributing significantly to the global waste crisis (SCMP, 2020). During the first quarter of 2020, the United Nations reported the distribution of over 1 million gowns, 1.8 million surgical masks, and 6.4 million gloves worldwide. Projections indicated that low- and middle-income countries would require vast quantities of PPE, further complicating waste management efforts (Grand View Research, 2020).

The pandemic has also led to a marked increase in household waste due to shifts in consumer behavior, including online shopping and increased food consumption during lockdowns (Sharma et al., 2020). The improper disposal of masks and PPE has raised serious environmental concerns, as these items frequently end up in water bodies, contributing to plastic pollution. The hydrophobic coatings used in masks, primarily made from PP and other materials like polyacrylonitrile (PAN) and polyurethane (PUR), have significantly increased the plastic waste load (Singh et al., 2022).

According to WHO modeling, the demand for PPE was staggering, with estimates of 1.6 million goggles, 76 million laboratory gloves, and 89 million medical masks required monthly for COVID-19 responses (WHO, 2020). The Worldwide Fund for Nature (WWF) estimated that around 10 million masks could be released into the environment monthly, leading to soil and groundwater contamination from improper BMW disposal (Sharma et al., 2020).

Developing countries face unique challenges in waste management, often employing less effective disposal methods than their developed counterparts. Insecure landfills and the lack of protective equipment for waste handlers increase the risk of virus transmission (Sharma et al., 2020). BMW, which includes highly infectious materials, poses significant health risks, especially since asymptomatic individuals may produce waste contaminated with the virus. Improper disposal of such waste can endanger the lives of waste management workers, particularly in developing nations where these workers often lack sufficient PPE.

Moreover, panic buying during the pandemic has led to excessive storage of perishable food items, further increasing waste generation without proper cold storage facilities. Lockdowns have shifted consumer behavior toward online shopping, amplifying waste packaging (both paper and plastic). The need for focused attention on waste management has never been more critical, especially given the complexities and volume of waste generated during this global crisis.

9.3 Impact of COVID-19 on medical waste

The pandemic has severely impacted BMW management systems worldwide, resulting in an unusual spike in medical waste generation. For instance, China reported a six-fold increase in medical waste, reaching approximately 240 metric tons (Tassakka et al., 2020). In Barcelona, the volume of medical waste surged by 350%, with levels rising from about 275 tons to 1200 tons during the pandemic (ACR, 2020).

Governments are increasingly focused on managing waste with new systems and guidelines to effectively respond to the pandemic. The Central Pollution Control Board (CPCB) in India has issued revised guidelines on the handling, treatment, and disposal of waste generated during the treatment, diagnosis, or quarantine of COVID-19 patients (CPCB, 2021).

The climate change crisis has become a pressing global issue, and the COVID-19 pandemic in 2020 highlighted the need for action. Efforts to control COVID-19 transmission temporarily improved air quality in some regions, with global CO_2 emissions dropping by 17% compared to 2019 levels (Le Quéré et al., 2020). While there were temporary decreases in emissions, these changes are likely short-lived; as economies reopen, emissions may surge back to or exceed prepandemic levels. Although there is a decline in oil prices impacting the renewable energy market, evidence suggests that climate conditions do not significantly influence virus transmission (Downtoearth, 2021).

To address the challenges posed by COVID-19, specific guidelines for managing infectious waste and other contagious diseases, such as HIV and H1N1, have been developed by various health authorities, including the Ministry of Health and Family Welfare (MoHFW), WHO, and CDC (CPCB, 2020). India generated 18,000 tons of medical waste within 4 months during the pandemic, with the daily healthcare waste from a COVID-19 patient averaging 3.4 kg. This waste generation spiked during peak periods of the pandemic, with reports of a 600% increase in healthcare waste in regions like Hubei (Das et al., 2021).

European governments faced challenges managing household waste generated by COVID-19 patients, ensuring the safety of healthcare workers, and providing adequate space for waste disposal. The CPCB has outlined guidelines for the safe handling, treatment, and disposal of waste generated during COVID-19 patient care, emphasizing proper collection and disposal according to the BMW management rules,

2016 (CPCB-BMW, 2016). These guidelines detail the responsibilities of various stakeholders, including Common BMW Treatment Facilities (CBWTF) and urban local bodies.

The reverse logistics network design has been employed for COVID-19 waste management. Medical waste is initially collected from treatment centers and handled according to BMW management rules, 2016, and CPCB guidelines for COVID-19. The materials are then temporarily stored in designated centers before being disposed of using special landfill methods (Jebaranjitham et al., 2022). Proper sanitization protocols for collection vehicles and record-keeping for the collection, treatment, and disposal of COVID-19 waste are critical for ensuring effective waste management during this ongoing public health crisis.

9.4 Methods for biomedical waste disposal

The COVID-19 pandemic has posed considerable challenges for European recycling firms, as the reduced costs and increased safety of virgin plastics over recycled alternatives create significant economic pressures. This situation may lead to greater incineration of recyclable waste, negatively impacting the environment and reducing the recovery of essential plastic resources. Disposable surgical masks, essential during the pandemic, are produced from various polymeric materials—including PP, PUR, PAN, polystyrene (PS), polycarbonate, PE, and polyester—that are foundational for many plastic products. A single-use mask generally consists of three layers: A fibrous inner layer, a melt-blown central filter layer, and a water-resistant, colored outer layer. The central layer is often produced using electrospinning of nanofibers or microfibers, in which melted polymer components are charged and exposed to an electric field to optimize particle filtration (Aragaw, 2020).

Plastic degradation occurs via several pathways—biodegradation, photodegradation, chemical degradation, and thermal degradation. Biodegradation involves microorganisms like bacteria and fungi, which create biofilms on plastic surfaces, breaking down molecular structures to yield oligomers, dimers, and monomers via enzymatic activity. Photodegradation and chemical degradation facilitate biodegradation, as UV exposure degrades plastics into compounds usable by microorganisms. Thermal degradation has additional applications, including identifying plastic types in environmental samples through pyrolysis gas chromatography-mass spectrometry (Pyr-GC/MS) or generating syngas and fuel oils at high temperatures. This method can also be employed at lower temperatures as a pretreatment to make plastics more susceptible to biodegradation by altering their chemical structure (Liu et al., 2022). Chemical recycling processes such as gasification and pyrolysis hold promise for large-scale applications, generating fuels, chemicals, and hydrogen from plastic waste (Lopez et al., 2018). Each degradation method, however, presents challenges: strong carbon–carbon bonds can hinder biodegradation, photodegradation may produce toxic by-products, and advanced oxidation mechanisms remain under research. Additionally, while thermal degradation offers biofuel potential, it carries notable limitations

(Liu et al., 2022). Besides incineration and thermal degradation, several alternative technologies are available for the treatment of BMW, each with unique benefits. These alternatives include autoclaving, gas sterilization, chemical disinfection, microwave treatment, irradiation, and thermal inactivation. Recently, hydrothermal carbonization (HTC) has shown promise for the treatment of noninfectious plastic medical waste, using high-pressure and high-temperature autoclaving for effective carbonization. Selection among these technologies should consider factors such as affordability, adaptability, cost, and associated health and environmental impacts (Lenka et al., 2024).

Plastics' decomposition has significant environmental implications, as it releases greenhouse gases like methane and ethylene, contributing to climate change. Biodegradable plastics present a potential solution to these issues by offering a reduced environmental impact compared to conventional noncompostable plastics (Abbasi & Sıcakyüz, 2024). Traditional plastics are often derived from fossil-based sources, whereas biodegradable plastics utilize biological feedstocks, offering a more sustainable alternative . Combining multiple degradation methods could improve the breakdown of mixed plastic types, using light and heat to weaken specific chemical bonds, thus making them more amenable to biodegradation and resulting in low-molecular-weight polymers that are easier to process (Liu et al., 2022). Reusing plastics is a favorable alternative to recycling as it consumes less energy, reduces resource demand, and limits emissions, as plastic production accounts for 4%–8% of global oil consumption, both as feedstock and through manufacturing processes. Reuse also lowers MSW, as well as CO_2, nitrogen oxides, and sulfur dioxide emissions (Al-Salem et al., 2009). Additionally, plastics have demonstrated reusability in various applications, such as construction materials—replacing bricks or as aggregates in asphalt—illustrating their value in resource efficiency (Haque et al., 2021).

9.4.1 Autoclave treatment

Autoclave treatment uses high-temperature steam to sterilize or disinfect BMW, offering an accessible, eco-friendly method with relatively low operational costs. This process is efficient and does not require specialized operators, though its effectiveness can vary based on load size. Although the installation of autoclaves is costly, they are reusable and highly effective. A potential drawback is the release of high efficiency particulate air (HEPA) emissions, which require filtering to prevent toxic fumes and odors. Autoclaving is unsuitable for pharmaceutical and chemical wastes (Ojha et al., 2022).

9.4.2 Chemical treatment

Chemical treatment is a straightforward method that uses disinfectants like liquid chlorine, chlorine dioxide, sodium hypochlorite, ultraviolet light, and ozone to sterilize BMW in healthcare facilities. While effective, it generates effluents, notably chlorine gas, which can have adverse environmental effects. To optimize the process and minimize risk, BMW should be thoroughly shredded before chemical treatment (Wang et al., 2020).

9.4.3 Incineration

Incineration is a widely adopted method in healthcare facilities, often required by regulation. While effective for BMW disposal, this process is costly due to the need for secondary treatment facilities to mitigate harmful emissions, including dioxins and furans, which contribute to greenhouse gas emissions. Despite high installation costs, incineration remains the preferred choice for its high efficiency and capacity (Ojha et al., 2022).

9.4.4 Microwave treatment

Microwave treatment utilizes thermal depolymerization and reverse polymerization in an oxygen-free (N_2) environment, applying high-energy microwaves to break down organic BMW substances. Electromagnetic waves raise the internal energy of materials, promoting disintegration (Ojha et al., 2022).

9.4.5 Dry-heat treatment

A high-temperature sterilization technique, dry-heat treatment is best suited for dry materials like glassware, powders, and wrapped items. Using conduction, convection, and radiation, it provides noncorrosive sterilization ideal for metal items but unsuitable for plastics and rubber. Temperature and timing must be carefully managed, particularly with plastic BMW items, to avoid degradation (Ojha et al., 2022).

9.4.6 Recycling personal protective equipment kits into value-added products

Advanced processes, including glycolysis, hydrolysis, pyrolysis, aminolysis, gasification, thermochemical recycling, and hydrogenation, can convert PPE plastic scraps into valuable products. Pyrolysis and gasification yield products like petroleum gasoline and syngas, while other recycling methods produce biodiesel, biofuels, and plastic polymers, repurposable in commodities such as shoes and garden furniture (Lenka et al., 2024).

9.4.7 Physico-chemical methods

This method begins with collection and sorting based on material characteristics, followed by decontamination and shredding for further processing. Resizing helps identify metals or other elements in the waste. After resizing, plastics are sorted by density, melting point, color, and other properties, then processed through extrusion or compounding. This final step melts plastic fragments into granules, ready for innovative recycling uses (Lenka et al., 2024).

9.4.8 Pyrolysis

Pyrolysis involves the application of high temperatures, although at levels lower than those used in incineration. Several advanced pyrolysis techniques, including induction-based pyrolysis, pyrolysis oxidation, laser-based pyrolysis, and plasma pyrolysis, have been developed. This method is considered one of the safest and most efficient for waste disposal, as it achieves substantial mass reduction with minimal emissions. Importantly, pyrolysis chambers destroy toxic gaseous vapors during the process, releasing a clean exhaust stream known as syngas. This syngas can then be repurposed to generate electricity or for other fermentation applications (Ojha et al., 2022).

9.4.9 Encapsulation and inertization

Encapsulation and inertization involve crushing BMW into smaller pieces with crushers, followed by thorough disinfection. The treated waste fragments are then placed in metallic drums, sealed with plastic foam, and subsequently disposed of in landfills. Alternatively, the chunks are mixed with manufactured and river sand in three different ratios. The resulting material is tested for mechano-chemical properties such as compressive strength, tensile strength, and resistance, evaluating its potential as a sustainable substitute for traditional construction materials (Ojha et al., 2022).

9.5 Clustered regularly interspaced short palindromic repeats for waste management

The rapid advancement of high-throughput genomic sequencing technologies has revealed a multitude of genes linked to various biological processes, opening new avenues in bioremediation science. Environmental pollution remains a growing concern across soil, water, and air systems, posing substantial risks to human health and ecological balance. Today's pollution sources are largely anthropogenic, including mining operations, industrial waste disposal, fossil fuel consumption, and the extensive use of agrochemicals.

The COVID-19 pandemic has further complicated waste management challenges, especially with the rise in plastic waste. Public health priorities have shifted consumer behavior, resulting in increased use of PPE, a higher demand for plastic-packaged foods, and widespread reliance on disposable items. These trends have highlighted significant limitations in current waste management systems, with the mismanagement of plastic waste raising pressing environmental concerns as it accumulates in ecosystems.

Bioremediation is a process that leverages living organisms, such as bacteria, fungi, and plants, to break down harmful contaminants and transform them into less toxic substances. As a highly eco-friendly approach, it offers a sustainable solution to reduce environmental pollutants. In light of these challenges, genome editing technologies are proving to be transformative in the field of bioremediation. Researchers can now precisely alter gene expression to enhance three main bioremediation strategies: Microbial bioremediation, phytoremediation, and mycoremediation. A groundbreaking development has been the application of Clustered Regularly Interspaced Short Palindromic Repeats (CRISPR) technology, a powerful genomic tool that allows for the targeted improvement of specific traits in plants, bacteria, and fungi, thereby significantly boosting their capacity for pollutant removal (Sahoo et al., 2022). With advancements in science, researchers are focusing on enhancing bioremediation through genome editing technologies. These gene editing tools have shown great potential in improving processes like xenobiotic removal, the conversion of toxic compounds into safer forms, and the breakdown of pesticides into simpler components. Key gene editing technologies, including CRISPR-Cas, zinc finger nucleases (ZFN), and transcription activator-like effector nucleases (TALEN), hold promising applications for achieving these objectives (Sharma et al., 2024).

The development of phytoremediation technology has always been the focus of ensuring sustainable engineering genetic addition; with the advent of the industrial revolution, various pollution caused by wastewater and pollutants expanded simultaneously (Naz et al., 2022). Phytoremediation is an excellent nature-based technique. The technique is a contender based on plants' natural inclination to recover soil, sediments, and water. Following the current environmental picture, remediation is examined, and determined that phytoremediation is an environmentally benign, economically feasible, and sustainable disposal option. Moreover, more research is needed to move phytoremediation potential, focusing on applying numerous combinatorial techniques (Naz et al., 2022). Among genetic engineering tools, cluster repeats of spaced palindromic (CRISPR)-Cas9 have great potential for expression regulation at specific sites and accelerate the new understanding of plant functional genomics. The vital potential advantages of the CRISPR-Cas9 genome editing technique in phytoremediation have been explored recently (Naz et al., 2022). Using the CRISPR-Cas9 technique, the gene editing in plant growth-promoting rhizobacteria (PGPR) and plants can improve the synthesis of the bioactive compounds, which help increase biomass production, tolerance to pollutants, transportation, accumulation, and detoxification of critical pollutants (Naz et al., 2022).

9.6 Bioremediation: The network of biochemical processes

Bioremediation relies on various biochemical reactions that work together to break down contaminants, forming a complex network of metabolic pathways at polluted sites. Common bioremediation techniques include bioleaching, land farming, bioventing, bioreactors, bioslurping, bioaugmentation, composting, natural attenuation, phytoremediation, biostimulation, and rhizofiltration (Li et al., 2011).

Bioremediation at contaminated sites typically occurs through two main approaches. In the first, optimal conditions—such as nutrient availability, temperature, and oxygen levels—are maximized to stimulate the growth of indigenous microorganisms that naturally consume pollutants. Additionally, genetically modified microbes may be introduced to enhance degradation efficiency. Alternatively, contaminants may be transported to a secondary site, where they are processed based on the nature and size of the pollutants, and then exposed to microbes to continue bioremediation (Chakraborty et al., 2022).

This eco-friendly process harnesses a network of biochemical interactions designed to break down pollutants by expanding the metabolic activity of microorganisms through in situ or ex situ methods. Bioremediation can occur under both aerobic and anaerobic conditions, depending on the environmental context and contaminant type. In many cases, both modes may be applied individually or in combination. Certain bacterial species, such as *Pseudomonas, Sphingomonas, Alcaligenes, Mycobacterium*, and *Rhodococcus*, are known to degrade hydrocarbons and pesticides under aerobic conditions, using carbon as their primary energy source (Chakraborty et al., 2022).

Recent studies have documented a variety of fungi capable of degrading persistent plastic pollutants, isolated from diverse sources such as soils at plastic disposal sites, marine environments, agricultural mulch film waste, landfills, plant tissues, and insect guts. The biodegradation pathway typically begins with the production of hydrophobin and biofilm formation, followed by the secretion of specific plastic-degrading enzymes like peroxidases, hydrolases, proteases, and ureases (Khatua et al., 2024). These enzymes facilitate the penetration of the three-dimensional polymer structures, leading to the mineralization of plastic into nontoxic by-products. Various synthetic polymers, including PE, PS, PP, polyvinyl chloride, PUR, and certain biodegradable plastics, have been shown to degrade within months due to the activity of diverse fungal strains, particularly those from the Ascomycota phylum, such as *Alternaria, Aspergillus, Cladosporium, Fusarium*, and *Penicillium* species. Moving forward, research should focus on metagenomic approaches to better characterize polymer-degrading microbial communities and explore bioaugmentation strategies to enhance plastic waste management. The current accumulation of untreated plastic litter in terrestrial and aquatic ecosystems, from soil and freshwater bodies to oceans, has intensified the search for effective alternatives to plastic waste management. Biodegradation, being both sustainable and cost-effective, offers a promising solution by transforming plastic into water, carbon dioxide, and biomass with the help of living organisms, especially microbes (Anjana et al., 2020; Shah et al., 2008; Zeghal et al., 2021). Within this framework, fungi stand out for their ability to colonize hydrophobic plastic surfaces and secrete a variety of hydrolytic enzymes that break down the polymers, facilitating the transport of smaller fragments across cellular membranes for further intracellular degradation.

9.7 Editing microbial metabolic pathways

CRISPR technology is used to modify bacterial and fungal genomes to increase the expression of plastic-degrading enzymes. According to Zimmermann et al. (2024), CRISPR enhances plastic breakdown by knocking out competing metabolic pathways or introducing regulatory elements, ensuring microbial focus on efficient plastic degradation.

9.8 Multiplex gene editing in microbes

CRISPR also allows for the simultaneous editing of multiple genes within a single organism, enabling coordinated optimization of enzyme expression, secretion, and metabolic pathways. This multiplex approach is especially valuable for complex processes like Polyethylene terephthalate (PET) degradation, where enzymes such as PETase and MHETase need to function together. Through CRISPR, researchers have enhanced the thermostability of PETase by introducing targeted mutations in its gene, enabling it to operate effectively at higher temperatures and accelerating the PET degradation process (Phillip & Chauhan., 2024).

9.9 Successful applications of biotechnology in plastic degradation

With global plastic pollution reaching critical levels, breakthroughs in biotechnology—specifically in engineered enzymes and microbial solutions—are transforming how we manage plastic waste. From industrial innovations to field trials, several successful applications highlight the potential of these technologies to reduce environmental damage and support circular economy (CE) goals.

9.10 Carbios: Industrial scale PET recycling with engineered enzymes

The French biotech company Carbios has pioneered industrial applications of enzyme-based PET (PE terephthalate) recycling. Carbios developed a proprietary process utilizing engineered PETase enzymes to break down PET waste into its fundamental monomers, which are then reused to manufacture new PET products, establishing a closed-loop recycling system (Carbios & Indorama Ventures, 2022). Their enhanced PETase variant significantly improves efficiency on an industrial scale, enabling the enzymatic breakdown of PET into terephthalic acid (TPA) and ethylene glycol (EG), which are subsequently purified for new plastic production (Alex, 2022).

In 2021, Carbios launched a demonstration plant in France, capable of recycling thousands of tons of PET annually, showcasing the scalability of enzyme-based PET recycling for global applications. With plans to open the first commercial-scale PET recycling facility by 2025, Carbios exemplifies the feasibility of enzyme-driven plastic recycling on an industrial level, offering a more sustainable alternative to traditional recycling methods that often downcycle plastics into lower-quality materials (Michael, 2022). By creating a closed-loop system, Carbios supports a CE model, enabling continuous recycling of plastics into high-quality products and minimizing reliance on virgin plastics.

9.11 Academic innovations in engineered PETases

Academic research has been instrumental in advancing plastic degradation technologies, particularly in enhancing the effectiveness of PETases. Researchers at the University of Portsmouth, for instance, developed a "super-enzyme" by combining PETase with MHETase, significantly accelerating PET breakdown, a breakthrough with large-scale potential for recycling facilities and bioreactors (Oda & Wlodawer, 2024). By refining enzyme stability and activity to withstand industrial conditions like higher temperatures and varying pH levels, these engineered enzymes pave the way for bioreactor-based recycling systems, where plastics can be efficiently processed and recycled.

9.12 Deploying plastic-degrading bacteria in oceans and landfills

Pilot projects are underway to employ plastic-degrading bacteria in marine environments and landfill sites, targeting plastic pollution directly at the source. In marine settings, scientists are testing engineered bacteria equipped with enzymes like laccases and lipases, capable of breaking down plastics such as PE and PS floating in ocean waters (Phillip & Chauhan, 2024). Though in preliminary stages, these trials show potential in reducing microplastic pollution in coastal areas.

In landfill sites, bacteria modified to degrade plastics like PUR and PET are being tested under controlled conditions (Phillip & Chauhan, 2024). Strains such as *Pseudomonas putida* are engineered to metabolize these plastics into bio-based chemicals, which can then be harvested and reused, reducing plastic accumulation in landfills. These projects represent early but promising steps in using biotechnological advancements to address plastic waste in challenging environments.

Through innovative approaches across different environments, biotechnology is unlocking solutions that could reshape plastic waste management, with applications that span laboratory research, industrial deployment, and environmental remediation efforts.

9.13 Synthetic microbial consortia for plastic degradation

In synthetic biology, an emerging approach involves designing microbial consortia—collaborative groups of genetically engineered organisms that target the degradation of multiple plastic types (Adamu et al., 2023). For instance, a consortium pairing *Escherichia coli* engineered to express PETase with *P. putida* modified for PUR breakdown can simultaneously degrade both PET and PU plastics. By distributing tasks among different microbial species, this method enhances degradation efficiency (Adamu et al., 2023).

9.14 Comparative analysis: CRISPR-cas9 gene editing versus synthetic consortia in plastic degradation

Evaluating CRISPR-Cas9 gene editing technology alongside the synthetic consortia approach reveals distinct advantages for each in facilitating plastic degradation. CRISPR-Cas9 offers precise genome editing, enabling targeted enzyme optimization within individual microbes for specific plastic-degrading functions. Synthetic consortia, on the other hand, leverage the unique metabolic capabilities of multiple organisms, allowing them to tackle a wider range of plastics simultaneously. Together, these approaches represent complementary strategies, with CRISPR fine-tuning microbial pathways and synthetic consortia enhancing degradation through cooperative, multispecies interaction (Adamu et al., 2023).

9.14.1 Carbon fibers

Carbon fibers (CF) have become indispensable across numerous industrial sectors, primarily in composites for textiles, prepregs, PPE, automotive components, and various medical applications. These polymer-based fibers measure between 0.00020 and 0.00039 inches and contain approximately 92%–95% carbon by weight (Rubio-Romero et al., 2020). Common precursors include PAN and mesophase pitch, renowned for yielding fibers with remarkable tensile strength, high resistance to creep, chemical stability, low density, and unique properties like electrical conductivity and thermal resilience, especially when oxygen exposure is limited. Key production techniques for CF include pultrusion, filament or tape winding, liquid and injection molding, and vacuum bagging.

Thanks to these exceptional properties—thermal resistance, low density, elasticity, and robust strength—carbon fiber filaments have found applications across various industries, from building construction and railways to aerospace and marine vessels. They also play a significant role in manufacturing athletic equipment, storage solutions, defense materials, and more. With demand continuing to rise, it is projected that global carbon fiber demand will reach approximately 200,000 metric tons by 2025. The COVID-19 pandemic further accelerated the need for these fibers in the production of critical medical supplies like PPE and other healthcare tools.

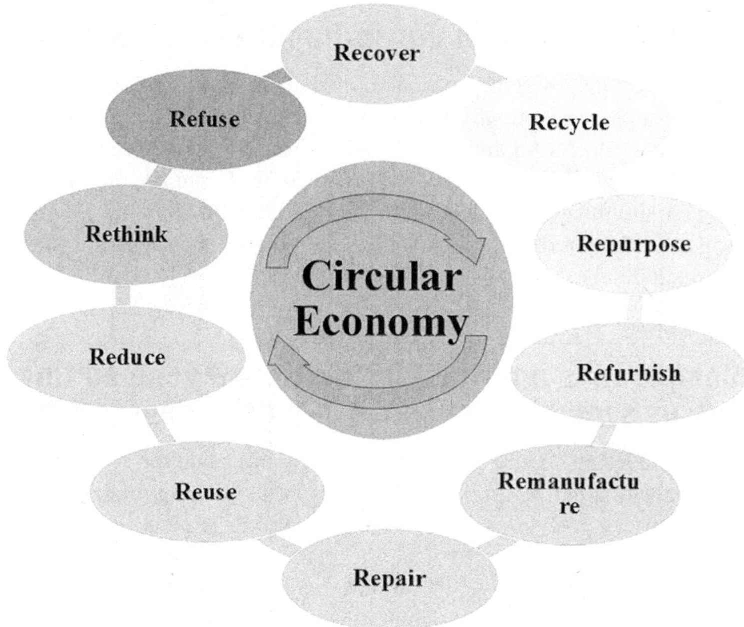

FIGURE 9.1

10R model of circular economy.

Given this increasing demand, managing the disposal of carbon fiber materials is becoming urgent. Proactive strategies to recycle and reuse these materials are essential to address environmental impacts effectively. While incineration and thermal decomposition can reduce waste volume, they present challenges. In addition to smoke, odor, and ongoing setup requirements, these processes release considerable heat, which complicates the recovery of usable CF. BMW, containing CF, such as disposable face masks, medical packaging, and hygiene products, often ends up in landfills, where burning releases toxic residues in ash and smoke.

A promising alternative to incineration involves processing biomedical carbon fiber waste at controlled temperatures and pressures, eliminating burning altogether. Stabilization—a controlled exothermic reaction that induces cyclization and oxidation of fibers at a low heating rate—could allow oxygen to diffuse fully into the precursor fibers, leading to complete stabilization. Recycled products from this process can be repurposed for various applications. Notably, Fitzer and Muller have demonstrated the cyclization process for PAN and its copolymers (Newcomb, 2016), indicating the potential for recycling biomedical carbon fiber waste into valuable materials with minimal environmental impact.

9.15 Recycling and recovery of carbon fiber

Current recycling processes in waste management often lead to downgraded material properties and value loss. However, an innovative protocol has been introduced to recover high-performance CF through thermal stabilization and pyrolysis, offering new possibilities for multifunctional use. For instance, fibrous filaments can be derived from discarded face masks, which are primarily made of PP. These materials can be converted into a carbon-rich, sulfur-doped, porous structure with a high carbon yield. The resulting CF possesses valuable properties suitable for various applications, such as oil adsorption and the removal of organic pollutants from marine environments. This method of recycling and reusing BMW offers a promising alternative to traditional waste management approaches, supporting economic sustainability by repurposing waste into high-value products. Biomedical accessories, such as personal protective face masks, have been recycled using twin-screw extrusion, yielding specialized compounds from comingled face masks. These compounds are then processed into pellets, which can be used as raw materials for creating new products, offering a sustainable path forward for BMW recycling (Mohanty et al., 2016; Yin et al., 2015).

9.16 The circular economy: A framework for sustainable waste management

The CE represents a paradigm shift in resource management, functioning as a regenerative system that aims to minimize resource consumption, waste generation, emissions, and energy loss through the strategic slowing, closing, and narrowing of material and energy loops. When integrated into waste management systems, CE principles serve as catalysts for economic recovery and growth, while simultaneously addressing the three pillars of sustainable development—economic prosperity, environmental stewardship, and social equity—through a synergistic approach that promotes energy recovery, material reclamation, and extended product lifecycles (Sharma et al., 2021).

This regenerative model's strength lies in its balanced approach, harmonizing economic performance with social inclusiveness and environmental resilience to benefit both present and future generations. The transition from current linear material flows to circular systems demands urgent policy interventions and comprehensive strategic planning. Central to this transformation is the "10R" framework, which delineates ten key strategies: recover, recycle, repurpose, remanufacture, refurbish, repair, reuse, reduce, rethink, and refuse (Fig. 9.1).

Current research in CE and waste management identifies five core themes:

1. Greenhouse gas reduction initiatives
2. CE integration with waste recycling
3. Life cycle analysis methodologies

Table 9.1 Modern biomedical waste management color coding system.

Container color	Contents	Specifications	Disposal
Yellow container: High-risk waste 	Includes medical PPE, clinical waste, expired pharmaceuticals, and chemical waste. Requires careful handling due to high contamination risk.	Leak-proof, heavy-duty nonchlorinated plastic with clear biohazard symbol. Ensures safe containment of hazardous materials.	High-temperature incineration or plasma pyrolysis. Anatomical waste may require secured deep burial.
Red container: Autoclavable waste 	Gather reusable PPE, plastic medical equipment, and laboratory consumables in red containers.	Process through steam sterilization at 121°C and 15 psi pressure.	Recycle suitable materials and recover reusable items when possible.
White container: Sharps management 	Accommodates hypodermic needles, surgical blades, and broken ampules. Ensures safe disposal of all sharp objects.	Puncture-resistant and tamper-evident design prevents accidental injuries. Essential for handling sharp medical items.	Undergoes autoclave sterilization or chemical disinfection. May require encapsulation in concrete for final disposal.

(*continued on next page*)

Table 9.1 Modern biomedical waste management color coding system—cont'd

Container color	Contents	Specifications	Disposal
Blue container: Glass and electronics 	Safely contains laboratory glassware and medicine vials. Prevents breakage and potential injuries.	Houses broken diagnostic equipment and mercury-containing devices. Ensures proper handling of electronic waste.	Undergoes chemical disinfection and specialized recycling processes. Ensures proper recovery of valuable materials.

Table 9.2 Key features of waste containers.

Symbol name	Symbols
Biohazard symbol (International biohazard symbol must be clearly displayed on all containers. Ensures immediate recognition of potential hazards.)	
Proper labeling (Containers labeled with waste type and collection date. Facilitates proper tracking and timely disposal.)	
Filling capacity (Maximum filling capacity set at 3/4 of container volume. Prevents overflow and ensures safe handling.)	

Table 9.3 Continuous improvement and compliance.

Weekly action	Monthly action	Quarterly action
Container sanitization	Staff refresher training	Comprehensive audit
Inventory check	Process efficiency review	Regulatory compliance check
Safety protocol review	Equipment maintenance	Stakeholder feedback analysis

4. Waste treatment innovations
5. Anaerobic digestion and resource recovery

Literature review findings indicate that CE discussions primarily center on recycling practices, followed by broader sustainable development principles. For optimal effectiveness, CE strategies should prioritize a hierarchy of actions: elimination, prevention, reuse, and recycling, with sustainable development serving as the foundational framework (Negrete-Cardoso et al., 2022).

According to BMW management rules, in 2016, the health centers must adopt a color-coded waste-parting system (Table 9.1, 9.2, and 9.3).

9.17 Conclusion and future prospects

The COVID-19 pandemic has underscored significant challenges within global waste management systems, revealing critical gaps in current strategies and spurring a renewed focus on sustainable waste solutions. Increased plastic waste generation, particularly from PPE and medical waste, has strained waste management infrastructure and highlighted the environmental risks posed by plastic accumulation, including the emergence of "plastispheres" in marine ecosystems. These developments underline an urgent need for innovative and integrative approaches to managing plastic and BMW. Key strategies, including biodegradation, thermal and chemical recycling, and advanced biotechnological interventions, hold promise for transforming waste streams into valuable resources, aligning with CE principles to close material loops and reduce environmental burdens.

Looking forward, the convergence of biotechnology, material science, and waste management offers transformative potential. Biotechnological advancements such as CRISPR-based microbial engineering and synthetic microbial consortia represent exciting possibilities for targeting complex plastic waste, enhancing biodegradation efficiency, and tailoring solutions to diverse environmental contexts. The adoption of multiplex gene editing in microbes and engineered enzymes (e.g., PETases) will be instrumental in enhancing the decomposition of high-molecular-weight plastics, contributing to a reduction in plastic pollution. Additionally, technologies like HTC, pyrolysis, and CRISPR-guided metabolic pathway modifications could improve waste processing efficiency, reduce greenhouse gas emissions, and enable the production of sustainable biofuels and high-value CF.

Future research must prioritize scalable bioremediation approaches and integrate CE principles into waste management practices. Policymakers and stakeholders should drive investment toward the development of sustainable infrastructure that supports recycling and bioconversion processes, fostering resilient waste systems that can withstand future public health crises. By committing to these pathways, we can lay the foundation for a sustainable waste management framework that not only mitigates current environmental impacts but also adapts to evolving ecological and societal needs.

References

Abbasi, S., & Sıcakyüz, Ç. (2024). A review of the COVID-19 pandemic's effects and challenges on worldwide waste management for sustainable development. *International Journal of Environmental Science and Technology*, 1–30.

ACR, 2020. Survey on the impact of the COVID-19 on municipal waste management systems. https://www.acrplus.org/en/municipal-waste-management-covid-19.

Adamu, K. S., Bichi, Y. H., Nasiru, A. Y., Babangida, A. M., Umar, M. M., Usman, G., & Muhammad, R. (2023). Synthetic microbial consortia in bioremediation and biodegradation. *International Journal of Research in Science, Innovation and Applied Sciences, 8*, 232–241.

Alex, T. (2022). Carbios mulls North American plant. *Chemical & Engineering News Global Enterprise, 100*(14), 10–11. doi:10.1021/cen-10014-buscon9.

Al-Salem, S. M., Lettieri, P., & Baeyens, J. (2009). Recycling and recovery routes of plastic solid waste (PSW): A review. *Waste Management, 29*, 2625–2643.

Anjana, K., Hinduja, M., Sujitha, K., & Dharani, G. (2020). Review on plastic wastes in marine environment–Biodegradation and biotechnological solutions. *Marine Pollution Bulletin, 150*, 110733.

Aragaw, T. A. (2020). Surgical face masks as a potential source for microplastic pollution in the COVID-19 scenario. *Marine Pollution Bulletin, 159*, 111517.

Barnes, D. K., Galgani, F., Thompson, R. C., & Barlaz, M. (2009). Accumulation and fragmentation of plastic debris in global environments. *Philosophical Transactions of the Royal Society B: Biological Sciences, 364*(1526), 1985–1998.

Bauchner, H., Fontanarosa, P. B., & Livingston, E. H. (2020). Conserving supply of personal protective equipment—A call for ideas. *Jama, 323*(19), 1911.

Carbios and Indorama Ventures. (2022). Carbios to build in France its first-of-a-kind manufacturing plant for fully biorecycled PET in partnership with Indorama Ventures. Press Release. https://www.carbios.com/en/carbios-to-build-in-france-its-first-of-a-kind-manufacturing-plant-for-fully-biorecycled-pet-in-partnership-with-indorama-ventures/

Chakraborty, R., Karmakar, S., & Ansar, W. (2022). Advances and applications of bioremediation: Network of omics, system biology, gene editing and nanotechnology. *Environmental Informatics: Challenges and Solutions*, 167–199.

CPCB, 2020. No revision 1: Guidelines for handling, treatment and disposal of waste generated during treatment/diagnosis/quarantine of COVID-19 patients central pollution control boardtitle web-page. https://www.mohfw.gov.in/pdf/63948609501585568987wastesguidelines.pdf.

CPCB, 2021. CPCB I central pollution control board web-page. https://cpcb.nic.in/covid-waste-management/.

CPCB-BMW, 2016. CPCB-BMW. https://www.cpcb.nic.in/uploads/Projects/BioMedical-Waste/BMW-GUIDELINES-COVID.pdf.

Das, A. K., Islam, M. N., Billah, M. M., & Sarker, A. (2021). COVID-19 pandemic and healthcare solid waste management strategy–A mini-review. *Science of the Total Environment, 778*, 146220.

Downtoearth, 2021. Last yr saw the biggest drop in CO2 emissions; here's why that's not enough web-page. https://www.downtoearth.org.in/news/pollution/last-yr-saw-thebiggest-drop-in-co2-emissions-here-s-why-that-s-not-enough-76745.

Gardiner, B. (2020). Pollution made COVID-19 worse. Now, lockdowns are clearing the air. *National Geographic*. https://www.nationalgeographic.com/science/2020/04/pollution-made-the-pandemic-worse-but-lockdowns-clean-the-sky/accessed.

Grand View Research. (2020). *Personal protective equipment market size, share, and trends analysis report by product (respiratory protection, protective clothing), by end use (Chemical, oil & gas, mining, construction), and segment forecasts*. Grand View Research.

Haque, M. S., Sharif, S., Masnoon, A., & Rashid, E. (2021). SARS-CoV-2 pandemic-induced PPE and single-use plastic waste generation scenario. *Waste Management & Research: The Journal of the International Solid Wastes and Public Cleansing Association, Iswa, 39*, 3–17. doi:10.1177/0734242X20980828.

Jebaranjitham, J. N., Christyraj, J. D., Prasannan, A., Rajagopalan, K., Chelladurai, K. S., & Gnanaraja, J. K. (2022). Current scenario of solid waste management techniques and challenges in Covid-19–A review. *Heliyon, 8*(7), e09855.

Kaza, S., Yao, L., Bhada-Tata, P., & Van Woerden, F. (2018). *What a waste 2.0: A global snapshot of solid waste management to 2050*. World Bank Publications.

Khatua, S., Simal-Gandara, J., & Acharya, K. (2024). Myco-remediation of plastic pollution: Current knowledge and future prospects. *Biodegradation, 35*(3), 249–279.

Klemeš, J. J., Van Fan, Y., Tan, R. R., & Jiang, P. (2020). Minimising the present and future plastic waste, energy and environmental footprints related to COVID-19. *Renewable and Sustainable Energy Reviews, 127*, 109883.

Lenka, J., Khuntia, S., Kar, B., & Sahoo, S. (2024). COVID-19 global biomedical waste, upcycling and management strategies. In *In Impact of COVID-19 waste on environmental pollution and its sustainable management: COVID-19 waste and its management* (pp. 145–169). Springer Nature Switzerland.

Le Quéré, C., Jackson, R. B., Jones, M. W., Smith, A. J., Abernethy, S., Andrew, R. M., De-Gol, A. J., Willis, D. R., Shan, Y., Canadell, J. G., & Friedlingstein, P. (2020). Temporary reduction in daily global CO_2 emissions during the COVID-19 forced confinement. *Nature Climate Change, 10*(7), 647–653.

Li, Y. Y., & Li, B. (2011). Study on fungi-bacteria consortium bioremediation of petroleum contaminated mangrove sediments amended with mixed biosurfactants. *Advanced Materials Research, 183*, 1163–1167.

Liu, L., Xu, M., Ye, Y., & Zhang, B. (2022). On the degradation of (micro) plastics: Degradation methods, influencing factors, environmental impacts. *Science of the Total Environment, 806*(Pt 3), 151312.

Lopez, G., Artetxe, M., Amutio, M., Alvarez, J., Bilbao, J., & Olazar, M. (2018). Recent advances in the gasification of waste plastics. A critical overview. *Renewable and Sustainable Energy Reviews, 82*, 576–596.

Mallapur, C. (2020). Sanitation workers at risk from discarded medical waste related to COVID-19. *IndiaSpend*. https://www.indiaspend.com/sanitation-workersat-risk-from-discarded-medical-waste-related-toCOVID-19/.

Michael, M. (2022). Carbios to build PET biorecycling plant. *Chemical & Engineering News Global Enterprise, 100*(8), 10. doi:10.1021/cen-10008-buscon12.

Mohanty, S., Ghosh, S., Nayak, S., & Das, A. P. (2016). Bioleaching of manganese by *Aspergillus oryzae* isolated from mining deposits. *Chemosphere, 172*, 302–309.

Naughton, C. C. (2020). Will the COVID-19 pandemic change waste generation and composition?: The need for more real-time waste management data and systems thinking. *Resources, Conservation, and Recycling, 162*, 105050.

Naz, M., Benavides-Mendoza, A., Tariq, M., Zhou, J., Wang, J., Qi, S., Dai, Z., & Du, D. (2022). CRISPR/Cas9 technology as an innovative approach to enhancing the phytoremediation: Concepts and implications. *Journal of Environmental Management, 323*, 116296.

Negrete-Cardoso, M., Rosano-Ortega, G., Álvarez-Aros, E. L., Tavera-Cortés, M. E., Vega-Lebrún, C. A., & Sánchez-Ruíz, F. J. (2022). Circular economy strategy and waste management: A bibliometric analysis in its contribution to sustainable development, toward a post-COVID-19 era. *Environmental Science and Pollution Research, 29*(41), 61729–61746.

Newcomb, B. A. (2016). Processing, structure, and properties of carbon fibers. *Composites Part A: Applied Science and Manufacturing, 91*, 262–282.

Oda, K., & Wlodawer, A. (2024). Development of enzyme-based approaches for recycling PET on an industrial scale. *Biochemistry, 63*(4), 369–401.

Ojha, P. C., Satpathy, S. S., Ojha, A. K., Sukla, L. B., & Pradhan, D. (2022). Overcoming challenges due to enhanced biomedical waste generation during COVID-19 pandemic. *Science of the Total Environment, 832*, 155072.

Phillip, A., & Chauhan, T. (2024). Innovative biotechnological approaches for plastic degradation: A pathway to sustainable waste management. *International Journal of Science and Research Archive, 13*(1), 2228–2243.

Pokhrel, S., & Chhetri, R. (2021). A literature review on impact of COVID-19 pandemic on teaching and learning. *Higher Education for the Future, 8*(1), 133–141.

Priyadarshini, B., Santhosh Kumar, M., & Das, A. P. (2024). Microbial degradation of synthetic plastics released from COVID-19 generated single-use disposable face masks. In *In impact of COVID-19 waste on environmental pollution and its sustainable management: COVID-19 waste and its management* (pp. 1–16). Springer Nature Switzerland.

Ranjbari, M., Esfandabadi, Z. S., Gautam, S., Ferraris, A., & Scagnelli, S. D. (2023). Waste management beyond the COVID-19 pandemic: Bibliometric and text mining analyses. *Gondwana Research, 114*, 124–137.

Rodić, L., & Wilson, D. C. (2017). Resolving governance issues to achieve priority sustainable development goals related to solid waste management in developing countries. *Sustainability, 9*(3), 404.

Rubio-Romero, J. C., Pardo-Ferreira, M. C., Torrecilla-García, J. A., & Calero-Castro, S. (2020). Disposable masks: Disinfection and sterilization for reuse, and non-certified manufacturing, in the face of shortages during the Covid-19 pandemic. *Safety Science, 129*, 104830. doi:10.1016/j.ssci.

Sahoo, S., Routray, S. P., Lenka, S., Bhuyan, R., & Mohanty, J. N. (2022). CRISPR/Cas-mediated functional gene editing for improvement in bioremediation: An emerging strategy. In *Omics insights in environmental bioremediation* (pp. 635–664). Springer Nature Singapore.

SCMP (South China Morning Post). (2020). China boosts face mask production capacity by 450 percent in a month, threatening a glut scenario. *SCMP*. https://www.scmp.com/business/companies/article/3075289/china-boosts-facemask-production-capacity-450-centmonth?module=perpetual_scroll&pgtype=article&campaign=3075289.

Shah, A. A., Hasan, F., Hameed, A., & Ahmed, S. (2008). Biological degradation of plastics: A comprehensive review. *Biotechnology Advances, 26*(3), 246–265.

Sharma, H. B., Vanapalli, K. R., Cheela, V. S., Ranjan, V. P., Jaglan, A. K., Dubey, B., Goel, S., & Bhattacharya, J. (2020). Challenges, opportunities, and innovations for effective solid waste management during and post-COVID-19 pandemic. *Resources, Conservation and Recycling, 162*, 105052.

Sharma, H. B., Vanapalli, K. R., Samal, B., Cheela, V. S., Dubey, B. K., & Bhattacharya, J. (2021). Circular economy approach in solid waste management system to achieve UN-SDGs: Solutions for post-COVID recovery. *Science of the Total Environment, 800*, 149605.

Sharma, V., Sehgal, R., Angra, V., & Gupta, R. (2024). Applications of genome editing in bioremediation. In *Genome editing in bacteria (Part 2)* (pp. 150–183). Bentham Science Publishers.

Singh, E., Kumar, A., Mishra, R., & Kumar, S. (2022). Solid waste management during COVID-19 pandemic: Recovery techniques and responses. *Chemosphere, 288*, 132451.

Tassakka, M. I., Musrianton, M., Alsita, I., Runtu, K. G., & Normayasari, N. D. H. (2020). Modeling the impact of the COVID-19 pandemic on air emissions indicators of climate change originating from solid waste management in coastal settlements. *IOP Conference Series: Earth and Environmental Science, 584*(1), 012062.

UNEP. (2020). BASEL: Waste management an essential public service in the fight to beat COVID-19. *The United Nations Environment Programme (UNEP) and the Basel Convention* http://www.basel.int/Implementation/PublicAwareness/Press%20Releases/WastemanagementandCOVID19/tabid/8376/Default.aspx.

Wang, J., Shen, J., Ye, D., Yan, X., Zhang, Y., Yang, W., Li, X., Wang, J., Zhang, L., & Pan, L. (2020). Disinfection technology of hospital wastes and wastewater: Suggestions for disinfection strategy during coronavirus disease 2019 (COVID-19) pandemic in China. *Environmental Pollution*, 114665.

WHO. (2020). *Shortage of personal protective equipment endangering health workers worldwide*. The World Health Organization https://www.who.int/news-room/detail/03-03-2020-shortage-of-personalprotective-equipment-endangeringhealthworkers-worldwide.

Wilson, D. C., Rodic, L., Modak, P., Soos, R., Carpintero, A., Velis, K., Iyer, M., & Simonett, O. (2015). *Global waste management outlook*. UNEP.

Yin, S., Tuladhar, R., Shi, F., Shanks, R. A., Combe, M., & Collister, T. (2015). Mechanical reprocessing of polyolefin waste: A review. *Polymer Science and Engineering, 55*, 2899–2909. doi:10.1002/pen.24182.

Zeghal, E., Vaksmaa, A., Vielfaure, H., Boekhout, T., & Niemann, H. (2021). The potential role of marine fungi in plastic degradation–A review. *Frontiers in Marine Science, 8*, 738877.

Zimmermann, A., Prieto-Vivas, J. E., Voordeckers, K., Bi, C., & Verstrepen, K. J. (2024). Mutagenesis techniques for evolutionary engineering of microbes–exploiting CRISPR-Cas, oligonucleotides, recombinases, and polymerases. *Trends in Microbiology, 32*(9), 884–901.

CHAPTER

10 The relational governance of household biomedical waste produced in South African informal settlements in a post-COVID era

Herman Geyer

Department of Geography and Environmental Studies, Stellenbosch University, Stellenbosch, Western Cape, South Africa

10.1 Introduction

The COVID 19 pandemic highlighted the waste management challenges of household biomedical waste produced in informal settlements in the Global South (Geyer & van Lille, 2024). Illegally dumped household biomedical waste (also termed healthcare risk waste in South Africa), originating from informal settlements, forms an emerging health challenge because of its close association with the rapid incubation and transmission of local epidemics (Laouar et al., 2019). Because of the cost of healthcare services, the majority of healthcare items in informal settlements are self-administered at home and disposed of together with other hazardous materials in numerous inaccessible illegal dumpsites. Given the volume of healthcare waste and hazardous materials found in the household waste from informal settlements, combined with high population densities, poor water and sanitation systems, the prevalence of chronic health conditions and the rapid outbreak of various types of diseases, and a lack of healthcare options, household healthcare risk waste is a leading cause of periodic localized epidemics that regularly affect the vulnerable residents of informal settlements (Zikhathile & Atagana, 2018). Moreover, this is not a contained population, since workers from informal settlements generally tend to be more mobile and communicative in the manufacturing, retail, and service industries, where they can transmit diseases throughout the city.

Household healthcare risk waste is not easily treated in developing countries because of the high labor demands for the manual removal, sorting, and disposal of household waste, and the inadequacy of waste management and road infrastructure in informal settlements for implementing conventional methods of biomedical waste removal. Whereas the modular governance structure of conventional waste management reduces the impact of healthcare risk waste in household waste, these mechanisms

Biomedical Waste Management in the COVID-19 Era: A Global Scenario.
DOI: https://doi.org/10.1016/B978-0-443-21841-5.00004-7

are not applicable in informal settlements, particularly in the Global South, due to regulatory challenges. Therefore, the study evaluates relational governance structures employed in the study area to minimize the health impacts of biomedical waste produced in informal settlements. This is a study area with a large population living in informal settlements, and therefore, the strategies employed in the case study area are of particular relevance for minimizing the transmission of new epidemics originating in informal settlements.

The structural and logistical inadequacies of safely managing household healthcare risk waste produced in informal settlements are a significant challenge for researchers. The study, therefore, aims to evaluate these challenges by first presenting a literature review to better understand the potential health risks of illegally dumped household healthcare waste produced in informal settlements and the technical challenges in implementing conventional solid waste management in informal settlements, particularly biomedical waste procedures. Secondly, the study evaluates the status of household healthcare risk waste management in Cape Town. Finally, the study analyses three relational strategies employed to minimize the impact of household healthcare risk waste on health outcomes within the communities in informal settlements and efforts to improve waste management services by reducing the cost of the collection, sorting, and treatment of household healthcare risk waste. The three key strategies identified to minimize the impact of household healthcare risk waste include developing adaptive and flexible procedures to maintain regular waste services during frequent solid waste management emergencies; conditioning communities to adopt least-cost behaviors to improve waste management services in informal settlements; and implementing trans-jurisdictional coordination between different solid waste management functions within different authorities.

10.2 The healthcare challenges of household biomedical waste

Healthcare risk waste is a significant challenge to public health in the developing world, with an estimated 5 million fatalities linked to infections from poorly managed biomedical waste. Furthermore, exposure to environmental pollutants is a causal factor in a quarter of all illnesses globally (Singhdeo et al., 2024). The COVID-19 era highlighted the potential for the rapid transmission of disease from biomedical waste (Rahman et al., 2020). However, similar to waste produced in healthcare facilities, household waste in developing countries also contains significantly high percentages of biomedical waste because of the prevalence of self-medication and informal healthcare among the urban poor (Zikhathile & Atagana, 2018). This includes medical waste generated by infected persons, sharps, toxic chemicals, and pharmaceuticals. However, it also includes significant amounts of sanitary products, including diapers, sanitary pads, contraception items, tissue paper, and pet litter, which are no less harmful since it is potentially contaminated by bodily fluids from infected persons or animals. Another risk is that potential pathological waste, including human tissue and animal carcasses, is often disposed of with household waste, contributing to the distribution of foot-and-mouth disease, avian influenza, and transmissible spongiform encephalopathy (Chowdhury et al., 2019). Due to poor waste management in developing countries, persons living in informal settlements often come into contact with infectious pathogens and toxic chemicals, either through direct contact by children and informal waste workers, or through indirect contact with environmental toxins, shared reticulation and sanitation facilities, or through vermin such as flies, rats, and dogs.

Household waste, particularly in developing countries, often contains hazardous materials, including toxic chemicals, heavy metals, and airborne hazards such as asbestos building materials. Moreover, much of the toxicity is environmental, consumed through airborne toxins from smoke and vapor inhalation, and biomedical pollutants that are bioactive in aquatic environments and soils. Toxins and pathogens are also transmitted through contaminated soil and groundwater into the products of urban agriculture (Squire, 2013). Only a low percentage of biomedical waste in developing countries is produced by healthcare clinics because of the proportionally high costs of private healthcare, and solid waste dumps contain a significant proportion of biomedical waste, much of which is produced outside of healthcare facilities (Dieng et al., 2022).

The urban poor living in informal settlements have a higher prevalence of communicable diseases than in other locations, because of the regularity of direct contact with the waste, the high population densities, and frequent transmission through surfaces such as water taps, toilets, and playgrounds. Contagious diseases remain viable on surfaces for up to 72 h, and therefore, biomedical waste mixed with household waste is an important source of infection (Kashyap et al., 2020). These communities regularly experience local epidemics, including diphtheria, *Escherichia coli, Staphylococcus*, HIV, and hepatitis. The microbial loads are significantly higher in informal settlements, and the local populations also suffer from respiratory, dermatological, eye, and pulmonary disorders from environmental pollution (Egbenyah et al., 2021; Zikhathile & Atagana, 2018). Furthermore, during epidemics, local communities often develop psychosociological responses to the epidemic, resulting in the increased household production of biomedical waste, resulting in the increased prevalence of healthcare items found in household waste (Kashyap et al., 2020). Consequently, household healthcare risk waste represents an environmental justice issue: communities demand a disease-free environment as a basic human right. The communities affected by biomedical waste tend to be socially disadvantaged, consisting of nonwhite, low-income populations. Moreover, this disease burden also represents a financial burden and a loss of human resources due to increased morbidities (Zikhathile & Atagana, 2018).

Following environmentally sound biomedical waste management procedures as prescribed by the Basel and Rotterdam Conventions, the proper disposal of biomedical waste requires the adequate segregation of biomedical waste, storage in encapsulated puncture-proof containers, transportation in special vehicles, regular scheduled disinfection and decontamination procedures, and the autoclaving, burial, or incineration of wastes in specialized facilities (Agrawal & Singh, 2005). Epidemics significantly increase the amount of biomedical waste produced, both in healthcare facilities and in domestic households. Since the principal source of waste is domestic, the increased use of biomedical waste creates a potential conflict in sustainable waste management (Ambali et al., 2013).

In contrast to the well-established procedures in dealing with biomedical waste produced from health centers, it is impractical to treat household healthcare risk waste in the same manner due to the large quantities of waste produced in residential areas, particularly during epidemics, with conventional biomedical waste protocols only applicable to waste produced by households with identified infections (Kashyap et al., 2020). Consequently, guidelines for household biomedical waste collection focus on relational strategies such as awareness campaigns through social media and the separate storage of biomedical waste in appropriate color-coded bags, which should be quarantined for 3 days before disposal in common landfills (Di Maria et al., 2020; Shammi et al., 2021). Most of the challenges associated with household healthcare risk waste are organizational, linked to the poor containment of waste at the point of origin, a lack of adequate policies on household biomedical waste, organizational failures in the collection, storage, transportation and disposal of household waste, the improper training of workers on the proper handling of biomedical waste, the inadequate communication of protocols to communities and improper design of infrastructure (Hassan et al., 2018).

Relational governance refers to the process by which rules and norms are formed to guide collective actions. Relational governance is particularly important in waste management since it steers local communities to adopt sustainable waste management processes (Giglio et al., 2020). It defines the rules, roles, and functions of different actors, creating modes of waste production that align with the procedures in solid waste management. In this way, local communities are integrated into the waste management process, thereby becoming key role players in sustainable waste management. Although interactions create tensions regarding the roles, obligations, and access to resources, they also provide opportunities for negotiating and developing trust and mutual commitment, thereby reducing the possibility of opportunistic behavior (Farrokhi et al., 2020).

This is linked to place-based relationalities that prioritize practices that relate to the sociocultural concept of the space as defined by the community. This relational approach enables officials in solid waste management functions to improve the level of service and to improve the sustainability of outcomes. This may influence the organization of everyday waste collection routines, the location of facilities, and the design of waste management infrastructure. Shifting waste management from a transactional arrangement to a social responsibility places the onus on the community members to conform to these practices, conditioning community behavior that is more conscientious and leads to more sustainable waste management (Middha & Horne, 2024).

10.3 Waste management challenges in informal settlements

The informal spatial and organizational structure of informal settlements severely limits the ability of local authorities to implement conventional waste management practices in these settlements. These settlements typically have an organic layout with very narrow, undulating roads, which are primarily designed to serve pedestrian and light vehicular traffic, and are not suited for waste compaction trucks. There are rarely any open spaces for waste infrastructure, and even when space is allocated for the disposal of waste, invariably the space is not sufficient for an adequate design following international standards of public dumping sites (Rodseth et al., 2020). Informal settlements are often located in terrain not suited for motorized access by waste removal vehicles, including floodplains, mountainsides, and road and rail servitudes. These settlements lack suitable road infrastructure, which makes conventional motorized waste removal costly or impossible in certain cases. Finally, the limited regulatory control over waste limits the capacity of local authorities to provide adequate basic infrastructure and services for waste management, since the residents in informal settlements either cannot pay for basic utilities or are unwilling to pay for merit goods, disincentivizing local authorities from providing waste management services. Since these informal settlements are technically illegal, local authorities cannot employ land use management enforcement to regularize community waste practices, but it also abrogates local authorities from any real responsibility to provide adequate waste management services to these communities (Opoko et al., 2016).

The manner of waste disposal also varies in different informal settlements. In certain settlements, waste removal is removed from a central disposal site, which may be a formally designated waste receptacle or a nondesignated waste disposal site, which becomes the de facto waste disposal site through community convention (Bruce & Storey, 2010). These sites are generally located next to major thoroughfare roads where waste management officials can access the site. However, in many cases, there are no designated waste disposal sites, and waste is invariably deposited at the periphery of the community, or in stormwater drains and wetlands, where seasonal rains will eventually redeposit the waste downstream, away from the community. In other cases, there are no suitable dumping sites, and waste is either burned or buried (Ochieng, 2019).

In most cases, the local community is not willing to pay for waste removal, despite the often exceedingly low costs of cross-subsidized waste management services. This is because informal communities often consider waste removal a merit good that no one wants to pay for and thus represents a market failure in waste management. The illegal dumping of waste breeds further poor waste management practices, as other residents are incentivized to dump waste and litter in inappropriate locations (Nanda & Berruti, 2021). This represents an environmental and health hazard because of the prevalence of vermin, children, and informal waste recyclers in and around the dump sites (Haywood, 2021). Unfortunately, this constitutes an environmental crime and a health hazard in terms of the South African National Environmental Management Waste Act (Act 59 of 2008). Informal waste adversely affects the soil, water, and air quality in informal settlements and results in the increased transmission of disease in these communities (Aparcana, 2017).

Consequently, there are different standards for waste management in different informal communities based on the practicality and cost efficiency considerations in different locations (Department of Environmental Affairs, 2011). Although formal settlements are structured for the customary weekly waste removal through motorized waste disposal trucks, informal settlements require more frequent removals because of the unavailability of closed waste dispensers and the open dumping of waste in illegal dump sites located close to informal housing, often without containment in refuse bags. Furthermore, it places an increasing burden on local authorities in terms of costs and labor because more waste workers are required to manually clear low volumes of waste, yet are required to clear the waste more regularly (Yadav et al., 2015). It also results in social costs as discarded waste blocks streams and stormwater channels, resulting in local flooding and the further contamination of the human environment with heavy metals, chemicals, and pathogens.

Therefore, relational governance strategies are preferred in informal settlements. This includes providing public education on domestic sustainable waste management and integrating informal waste recycling by bin pickers into waste management efforts (Ochieng, 2019). An important aspect of providing waste management services in informal settlements is reframing waste as a potential community resource and an employment strategy for local workers. It also implies utilizing waste management as a municipal strategy to promote local improvements and access foreign development assistance (Kain et al., 2016).

10.4 Methodology

The research is based on semistructured interviews with various stakeholders involved in solid waste management and land use management functions in Cape Town. The research involved 28 semistructured interviews conducted in person and telephonically. The respondents included persons in the various solid waste management departments and agencies in the City of Cape Town Metropolitan Municipality and the Provincial Government of the Western Cape. Interviews were also held with municipal councilors, community representatives, and law enforcement officers in areas with significant informal settlements, and with private sector contractors involved in healthcare risk waste management practices.

10.5 The household healthcare risk waste challenge in informal settlements in Cape Town

The safe and efficient solid waste management of household waste produced in informal settlements is a significant challenge for the local authorities in Cape Town.

In a city of approximately 4.7 million residents, an estimated 19.4% of the population lives in various types of informal settlements and housing structures throughout the city (StatsSA, 2020). There are an estimated 270 registered informal settlements distributed throughout the city, although most of these informal settlements are concentrated in and around the historic apartheid-era townships in the impoverished central Cape Flats district. While approximately 11.7% of the population live in conventional concentrated shack settlements, an additional 6.9% of residents live in informal backyard structures primarily constructed in and around the formal state-subsidized RDP housing settlements, and an additional 0.8% of residents live in tents and other temporary or mobile settlements in more affluent neighborhoods (CoCT, 2023).

The proliferation of shack settlements, backyard structures, and tent villages in the city places a significant strain on the solid waste management functions of the city. While almost all formal neighborhoods are provided with free mobile wheelie bins for weekly refuse removal using motorized waste compaction trucks, households in registered informal settlements do not have sufficient road infrastructure or secure spaces to store mobile wheelie bins. Consequently, all registered informal structures in informal settlements are issued with high-density domestic refuse bags for the safe collection and disposal of household waste. This process places a significant strain on labor resources, because these high-density domestic refuse bags have to be manually distributed daily to each household, and the waste refuse bags have to be regularly collected from over 850 informal settlement waste collection points. While the local authorities have attempted to mechanize the waste removal process through the provision of refuse skips, these skips are regularly stolen and repurposed. Many of the local communities have requested the removal of the skips because of safety concerns regarding the prevalence of criminal activities in and around the skips. Moreover, the location of each informal waste dumping site requires negotiation and coordination with the community representatives of each informal settlement to ensure community compliance; otherwise, the communities create separate illegal dumping sites, which are often situated in inappropriate or inaccessible locations such as wetlands or stormwater channels.

A further estimated 26% of households in informal settlements do not receive regular solid waste management services in the form of the daily distribution and collection of domestic refuse bags (Asivikelane, 2021). These primarily consist of unregistered informal settlements developed because of land invasions, illegal backyard structures in existing RDP settlements, and temporary tent cities. These settlements cannot claim regular solid waste removal services because these settlements are still illegal settlements subject to eviction orders until the land occupation claims are judicially reviewed following an extensive legal process. Consequently, the authorities cannot legally provide regular waste removal services to these households because strict national procurement regulations limit the provision of solid waste management services to registered formal and informal housing units. Instead, the solid waste management services provide periodic general waste collection and removal services through the Expanded Public Works Program. However, these waste removal services are even more labor-intensive and costly for the local authorities than the collection from informal waste dumping sites, since it requires the workers to canvass each settlement and the surrounding areas for distributed waste and litter. While the residents in backyard structures often have access to the wheelie bins of the government-assisted RDP housing unit on which they are located, these wheelie bins often have insufficient capacity because the wheelie bins are designed to serve the purposes of a single household, whereas most RDP housing units have 2–4 backyard structures housing separate households. Moreover, because wheelie bins are only provided to properly fenced RDP housing units to secure the bin from theft according to municipal bylaws, numerous illegal backyard structures do not have access to wheelie bins or refuse bags.

The critical challenges in solid waste management in informal settlements are accentuated by the increasing awareness of biomedical waste in household waste. Although biomedical waste is not a significant problem in conventional household waste because of the small volume of household healthcare risk waste, the mechanization of waste removal and the low level of morbidities, this is not so for the household waste produced in informal settlements. The urban poor in informal settlements have very high comorbidities, linked to prevalent chronic health conditions such as HIV, diabetes, and hepatitis, intestinal protozoa, helminth parasites, and chronic malnutrition. Combined with a high rate of self-medication, communal reticulation and sanitation facilities, poor diets, and frequent direct and indirect exposure to solid waste, these communities experience frequent localized epidemics. Recent epidemics in these settlements included tuberculosis, foot-and-mouth disease, listeriosis, rubella, staphylococcus, diphtheria, mumps, *Haemophilus influenzae*, and pertussis.

The recent COVID-19 pandemic highlighted the epidemic transmission risks of household healthcare waste produced in informal settlements. The combination of a highly vulnerable population, regular epidemics, and rapid disease transmission within the informal community makes the adequate management of household health risk a priority for solid waste management authorities. Since the residents in informal communities work throughout the city as temporary day laborers, and in the childcare, food, hospitality, and cleaning service industries, the containment of infectious disease transmission in informal settlements is an important objective in limiting the rapid incubation and distribution of epidemics.

The major challenge with household healthcare risk waste produced in informal settlements is the unconventional types of healthcare waste and the diverse mechanisms through which both waste and diseases are transmitted within these settlements. Because of the prevalence of chronic diseases and self-medication, much of household waste from informal settlements contains significant amounts of infectious waste, including needles, blades, used pharmaceuticals, and bodily fluids. A major challenge is used diapers, which form a significant percentage of household waste items, often containing infected fecal matter from sick children. These communities also produce significant amounts of animal carcasses for food and religious ceremonies because South African law permits the unregulated slaughter of animals for noncommercial purposes, because of its association with customary practices. Moreover, the re-utilization of discarded used cigarette nibs, which constitute approximately 18% of all litter, is a frequent source of oral infections. The local communities have a significantly higher level of direct and indirect exposure to household healthcare risk waste because of the illegal dumping of household waste in open sites, and the further distribution of waste by high winds, scavenger animals, children playing in dump sites, and informal waste recyclers. Since many of these informal settlements are built in floodplains unsuitable for formal housing development, they experience regular flooding events because most of the wetlands and stormwater infrastructure are blocked with discarded household waste.

10.6 The challenges of adapting conventional solid waste removal to informal household healthcare risk waste

The critical shortage of adequate solid waste removal services to informal settlements and backyard structures creates a significant health risk challenge in South Africa. According to the City of Cape Town's 2020 Integrated Waste Management Policy and the Western Cape Health Care Waste Management Act, 2007 (Act no. 7 of 2007), all waste that is potentially contaminated by pathogens must be treated as biomedical waste because it is potentially hazardous for the environment and human health. Moreover, these regulations require specialized collection, treatment, and incineration at the privately owned Vissershok Waste Management Facility, the only accredited biomedical waste disposal facility in the province (CoCT, 2020).

The collection and incineration of large volumes of household healthcare risk waste produced in informal settlements, following international biomedical waste treatment procedures, was deemed cost-prohibitive because of the large volume of waste produced in these settlements. Although household waste produced in informal settlements only constitutes approximately 3% of the total volume of waste produced in the municipality, it greatly exceeds the capacity of the Vissershok Waste Management Facility. Moreover, since the treatment of biomedical waste is highly specialized, the municipality subcontracts these services to accredited specialists in the private sector. In contrast, waste management in informal settlements is primarily accomplished through the manual removal of waste by unskilled temporary labor in the Expanded Public Works Program. These workers are typically not trained or equipped to collect and transport biomedical waste. The treatment of household healthcare risk waste as biomedical waste was deemed cost-prohibitive to the already overburdened solid waste management functions of local authorities, since it would require the retraining and equipping of most of the solid waste management labor force.

The household healthcare risk waste produced in informal settlements is not suitable for processing in a similar manner to biomedical waste. The waste produced in informal settlements is generally mixed, containing various quantities of household waste, construction and demolition waste, commercial and industrial waste, and organic refuse, and therefore it needs to be manually sorted in a manner that is not suitable for treating biomedical waste. Biomedical waste requires containment and quarantine protocols, limiting the physical handling of waste. Furthermore, biomedical waste is produced in a limited number of medical facilities with suitable waste management infrastructure, whereas household healthcare risk waste is produced in less accessible sites in numerous informal settlements without suitable waste management infrastructure.

However, the manual collection, transportation, sorting, and treatment of household waste that contains microbial communities and toxic chemicals also exposes solid waste management workers to communicable diseases. Therefore, specific protocols must be developed to reduce the risk of household healthcare waste produced in informal settlements. Since it is not possible to implement the rigorous procedures of biomedical waste management in informal settlements, the local authorities implemented several relational governance strategies to minimize the risk of household healthcare waste in informal settlements.

10.7 Flexible and adaptive procedures

The first relational governance strategy in dealing with household healthcare risk waste was developing flexible and adaptive procedures to deal with emerging solid waste management emergencies. The authorities recognized that the most important factor in limiting the rate of infection among the local community is maintaining uninterrupted waste management services, particularly in maintaining existing waste removal and clearance schedules. Recent experiences during the COVID-19 pandemic and lockdown emphasized that a single delay in the weekly waste removal schedule has an exponential effect on the number of actions that must be performed to clear waste removal backlogs. Service delays resulted in local skips becoming overfilled, resulting in domestic waste refuse bags being dumped around the skip, blocking access to skip removal trucks. Consequently, the skip removal contractors could not remove the skips and had to coordinate additional waste removal services to remove the refuse bags blocking the skips. After two weeks, many of the refuse bags degenerated in the high temperatures and winds, resulting in the accumulation of waste accumulating in local wetlands. Furthermore, there was a significant increase in rats and flies in the community, potentially increasing the rate of disease transmission.

Flexible and adaptive procedures require the rapid diversion of resources to a solid waste management emergency area to contain and ameliorate the effects of unexpected emergencies while maintaining the regular waste removal schedule. Recent solid waste management emergencies were as diverse as:

- a localized personnel shortage at a certain facility because of numerous infections;
- a local service protest in which the local community burned down the contractor's refuse removal truck;

- the overnight development of a new informal settlement through an organized land invasion of a power line servitude;
- a local gang war between paratransit syndicates, resulting in the refuse removal contractor refusing to provide waste removal services to the targeted community;
- a local listeriosis outbreak from tainted processed meat sold in an informal settlement, resulting in numerous hospitalizations and a local fatality.

While these events are nonlinear and unexpected, the local authorities could promptly address these waste management challenges by temporarily reducing services in other areas to skeleton staff levels and diverting the surplus labor and equipment resources to the affected area. During security challenges, specialized municipal law enforcement personnel, particularly the well-armed antiland invasion unit, were organized to protect the waste management services personnel in the performance of their duties. Most important, however, was the coordination with local clinics and hospitals to identify the location of an emerging epidemic and to treat the household waste at that location and adjacent locations as biomedical waste following established biomedical protocols and standard operating procedures.

Regarding health emergencies, such as emerging epidemics, specific procedures were developed and perfected during the COVID-19 era to address localized outbreaks. This included sending personnel into the community to inform the community representatives of the health emergency procedures and providing additional bags to each household to double-bag items that can be regarded as healthcare risk waste. It also required the rapid and frequent decontamination of shared water taps and toilet facilities, since these facilities were one of the primary sources of disease transmission. In certain cases, it was necessary to provide water tanker trucks and portable toilets to provide safe water reticulation and sanitation facilities to the affected community. Infected households were also identified and provided with additional resources to limit the spread of infections. Finally, procedures were implemented for the rapid removal and disposal of household waste and litter at Vissershok Waste Management Facility. The declaration of a solid waste management emergency also provides an ideal opportunity for the retraining and re-equipping of personnel. This includes the rapid acquisition of specialized equipment to address the specific waste issue and the decontamination of facilities and equipment. It also conditions personnel to process high volumes of potential biomedical waste, and it trains personnel to be more vigilant in terms of risky personal behaviors and practices.

However, developing flexible and adaptive procedures is not a simple process. Numerous legal and policy prescriptions impact solid waste management, not only within the domain of solid waste management but also in ancillary domains such as health and safety regulations, labor regulations, procurement protocols, budgetary procedures, and environmental guidelines, et cetera. Therefore, rapid responses to waste management emergencies require diverse and dynamic teams that can rapidly connect to a range of public and private sector actors within and outside the solid waste management field to address these challenges. Most importantly, the teams need to have access to emergency funds to rapidly acquire goods and services, particularly private sector contractors and approvals for overtime bonuses to address an emerging epidemic.

The two biggest challenges that prevent the development of flexible and adaptive procedures are overly risk-averse approaches to the health and safety risks of solid waste management personnel and the inability to compromise regarding competing legal and policy prescriptions. Risk-averse behavior during waste management emergencies often leads to service delays, extending the period of exposure. Similarly, it was found necessary to compromise and prioritize between competing policy prescriptions. For instance, procurement regulations and budgetary protocols needed flexibility during emergencies. Labor regulations also required flexibility in terms of job prescriptions, such as when a skip removal truck operator is required to move a couple of refuse bags to access the skip, or law enforcement personnel are required to periodically work after hours to contain illegal dumping or protect waste management personnel in emergency areas.

10.8 Community conditioning of least-cost behaviors

The second relational governance strategy in dealing with household healthcare risk waste was conditioning favorable community behaviors and practices as long-term conventions, which enable effective solid waste management in informal settlements. This requires a combination of incentives and disincentives that would result in inducing least-cost behaviors, i.e., induced behaviors that reduce solid waste management labor and operational costs. It was recognized that it is nearly impossible to force the residents in informal settlements to regularize their waste disposal behaviors to conform to all solid waste bylaw regulations because of the poverty of the community and the inadequacy of the waste-related infrastructure in informal settlements. However, it was possible to incentivize residents in informal settlements to dispose of their waste in a more orderly manner, which reduces the transmission of disease and makes it easier and more cost-effective to collect waste. The rationale of this strategy was to condition the residents to adopt and self-reinforce the safe and healthy disposal of household waste and to unlearn bad behaviors developed during the COVID-19 era when residents haphazardly dumped waste at the nearest street corner or wetland instead of in designated waste disposal sites.

To incentivize the safe and healthy disposal of household waste, the local waste management authorities apply a carrot-and-stick approach, by firstly providing information campaigns to make the local community aware of identified risk behaviors. During the COVID-19 lockdown, the community was informed regarding the separate double-bagging and the 2-day waste quarantining procedure for healthcare risk waste before disposal through flyers distributed to the individual structures and printed on posters displayed at public facilities. The local authorities also held information sessions regarding waste management challenges with community representatives and the local councilors representing the community, to relay the information back to the community during the weekly neighborhood association meetings. During solid waste management emergencies, these representatives were mobilized to rapidly relay emergency protocols to the various communities through local social media channels.

Secondly, the local waste management authorities engaged in negotiations with local community representatives to discuss practical solutions to informal household waste, which would both improve the health and safety of the community and reduce waste management labor and operational costs. The primary challenge is to identify the motivations for the illegal dumping of household waste in nondesignated sites and to address waste management concerns with local community representatives to ensure community compliance with solid waste management efforts. While many of the community's waste disposal behaviors seem irrational and counter-productive to health and safety concerns, upon further negotiation, it was often found that these were the most reasonable solutions given the challenges associated with living in inadequate housing conditions. Through negotiations with the community, alternative solutions could be developed that benefit both the community and waste management efforts.

For example, one of the major challenges is an unwillingness among residents in informal settlements to dump nondomestic waste in designated municipal recycling centers. This is linked to the high costs of transporting waste to the municipal recycling centers, which are often located far from informal settlements and are only accessible by vehicle. Consequently, informal settlements dumped items such as construction waste, commercial and industrial waste, and organic garden refuse in domestic waste skips designated for household waste, assuming that the municipality would remove and sort the waste. Another challenge raised was the safety issues associated with dumping household waste in designated disposal sites with low community visibility, where criminal activities occur, often associated with informal waste recyclers. A further challenge is a lack of awareness of the health and environmental impacts of illegal dumping of waste and an unconscientious attitude.

Some outcomes from these negotiations included the identification of suitable alternative household waste disposal sites located closer to specific communities with good visibility, the provision or removal of skips based on community health or security concerns, and the provision of suitable alternative disposal sites for bulky nonstandard household waste items such as discarded furniture, appliances, and construction waste in separate locations from domestic waste. Once an agreement was concluded with the community representatives, these representatives would place social pressure on the community to conform to the agreed actions.

Thirdly, it was important to improve the implementation of solid waste management services in informal settlements. This is especially important in unregistered informal settlements where there is an insufficient provision of waste management items, such as wheelie bins and refuse bags. As the respondents noted, if the informal settlements were regularly cleared of litter and household waste, the communities would become more accustomed to maintaining a clean environment by utilizing identified communal waste disposal sites. While national procurement policies prohibit the distribution of refuse bags to unregistered structures, the municipality implemented daily cleaning efforts to remove litter within and around the settlement and provided daily waste removals from identified waste disposal sites, which were paved and painted to demarcate the waste disposal site.

Finally, the local waste management authorities had to implement law enforcement to police egregious contraventions of solid waste management procedures. These included regular policing by law enforcement officials who implemented spot fines, the seizure of vehicles, and arrests for serious violations, while turning a blind eye to minor violations. For instance, the dumping of commercial and industrial waste by informal waste removers, the burning of waste to obtain recycled iron and copper and the dumping of waste in wetlands or stormwater channels were severely penalized, however, law enforcement officials permitted communities to dump household waste in an identified community waste collection point, even if that collection point was not designated as a registered waste disposal site since it reduced the labor costs of collecting waste in inaccessible locations. Informal waste recyclers were particularly targeted to prevent the hoarding of recyclable materials in less accessible sites.

10.9 Trans-jurisdictional coordination control

The third relational governance strategy in dealing with household healthcare risk waste was improving coordination and control between the different solid waste management functions of the local authorities. The fragmentation of waste management functions between different local authorities, departments, agencies, and external contractors was a significant impediment to effective healthcare risk waste management. In Cape Town, solid waste management functions are distributed between five different units in the municipal government and three different units in the provincial government. This segregated waste management responsibilities between different spaces within the same location to different authorities. For instance, the recreation parks department was responsible for waste removal in parks, while waste removal in the adjacent roads was the responsibility of the roads and transportation department. Moreover, only the law enforcement department had the authority to implement fines, seizures, and arrests. The fragmentation of waste management functions resulted in the duplication of waste removal efforts in the same location and a plausible deniability and blame shifting for poor solid waste management performance.

Since the complex bureaucratic structure of waste management services cannot be rationalized and simplified because of the complex nature of solid waste management, it is important to coordinate various waste management responsibilities. This requires clear and transparent communication channels and the integrated management of the various solid waste management functions. Transparent communications require direct personal contact between different waste management officials to coordinate waste management services. For example, it was necessary to publish a list of personal contact numbers to access different officials in the waste management services, since email ticketing processes often took days to resolve and often went unanswered when the central communications office forwarded the query to the wrong department.

Similarly, the officials could often not be accessed through their office numbers because they were in the field. The integrated management of the different solid waste management functions is also necessary so that the waste management responsibilities can be delegated through existing channels among the different sections of waste management services.

While this challenge has not been successfully resolved because of persistent disagreements between different authorities and departments and a reluctance of officials to provide open communication channels, a set of operational guidelines was established to enable the delegation of duties and to provide set communication channels through different mediums of communication between the different authorities, departments, agencies, and contractors. This included the use of email tickets with prescribed response times, call centers, and performance management guidelines. Unfortunately, it was not possible to centralize the overall coordination and management of waste services since many of the aforementioned departments and agencies did not have waste management as the central function, and the Municipal Structures Act (Act 117 of 1998) legislatively prescribes a decentralized governance structure.

However, progress was achieved by setting up a protocol for prioritizing continuous service delivery during waste management emergencies, since waste management was prioritized as an essential service following the Covid era. This included the development of backup plans and transitory protocols to reduce operations to skeleton staffing when it was deemed necessary to redirect resources to emergency areas. The budgetary structure of the departments was also organized in a manner enabling the solid waste management authorities to access funds for rapid implementation of smaller purchases and appointing approved external contractors during emergencies, bypassing some of the more onerous and lengthy procurement protocols. An important aspect of this is the scheduled evaluation of the integrity of equipment, particularly personal protective equipment, and the regular decontamination of waste management equipment.

10.10 Conclusions

The study evaluated the efficient management of household biomedical waste produced in informal settlements in Cape Town. Household biomedical waste, locally termed household healthcare risk waste, is a significant waste management challenge because an estimated 19.4% of the city's population lives in informal settlements, backyard housing, and tent villages without proper roads and waste infrastructure. Household biomedical waste is a health concern because of prevalent chronic health conditions, shared reticulation and sanitation facilities, frequent exposure to open waste, and high rates of self-medication, resulting in frequent local epidemics that are often transmitted to other areas of the city by workers.

Approximately a quarter of these households are unregistered and do not receive refuse bags because of national procurement regulations. Waste management is consequently highly costly and labor-intensive. Moreover, household biomedical waste contains unconventional items, creating additional health concerns. Unfortunately, household biomedical waste cannot be treated using the biomedical waste management protocols of healthcare facilities because of the high volumes of waste, the requirement of specially trained staff, and the mixed nature of waste produced in informal settlements, containing construction, commercial, and organic refuse and requiring manual removal, sorting, and disposal. Consequently, it is advised to reduce the health and safety impacts of household biomedical waste through relational governance strategies, which incentivize efficient organizational and behavioral solutions instead of modular governance strategies linked to technical biomedical waste management protocols.

Three specific relational governance strategies were implemented to minimize the impacts of household biomedical waste produced in informal settlements. The first was developing flexible and adaptive approaches to maintain uninterrupted waste management clearance schedules. This requires the rapid diversion of resources to various solid waste management emergencies and the coordination with local healthcare facilities to identify emerging epidemics. This also required coordination with community representatives, providing additional refuse bags to double-bag healthcare risk waste items, and the decontamination of shared reticulation and sanitation facilities. It also required rapid access to funds and networking with a range of actors outside of solid waste management.

The second relational governance strategy was the conditioning of favorable community behaviors that would reduce health and safety risks and reduce labor and operational costs. This included a combination of information campaigns to educate residents of the risks of biomedical waste, negotiations with community representatives to find practical solutions to waste management, improved solid waste management services to condition local residents to a waste-free environment, and strict law enforcement to regulate egregious contraventions of waste management bylaws.

The third strategy was improving the coordination between the complex and fragmented solid waste management functions in different local authorities, departments, agencies, and contractors to reduce the duplication of responsibilities and increase the accountability of service providers. This required transparent direct communication channels and the integrated management of the various waste management functions through a protocol prioritizing continuous waste management service delivery.

References

Agrawal, A., & Singh, R. (2005). *Understanding and simplifying bio-medical waste management*. Toxics Link Pamphlet.

Ambali, A. R., Bakar, A. N., & Merican, F. M. (2013). Environmental policy in Malaysia: Biomedical waste, strategies and issues. *Journal of Administrative Science, 10*(1), 1–17.

Aparcana, S. (2017). Approaches to formalization of the informal waste sector into municipal solid waste management systems in low-and middle-income countries: Review of barriers and success factors. *Waste Management, 6*(1), 593–607.

Asivikelane (2021). *Survey of cape town informal settlements.* International Budget Partnership.

Bruce, A., & Storey, D. (2010). Networks of waste: Informal economic systems and sustainability in Bali, Indonesia. *Local Economy, 25*(3), 176–189.

Chowdhury, S., Kim, G., Bolan, N., & Longhurst, P. (2019). A critical review on risk evaluation and hazardous management in carcass burial. *Process Safety and Environmental Protection, 123*, 272–288.

CoCT. (2023). *State of cape town report 2022.* City of Cape Town Metropolitan Municipality.

City of Cape Town (2020). *Five-Year Integrated Development Plan (July 2017-June 2022).* Retrieved from https://resource.capetown.gov.za.

Department of Environmental Affairs. (2011). *National waste management strategy.* DEA.

Di Maria, F., Beccaloni, E., Bonadonna, L., Cini, C., Confalonieri, E., La Rosa, G., Milana, M. R., Testai, E., & Scaini, F. (2020). Minimization of spreading of SARS-CoV-2 via household waste produced by subjects affected by COVID-19 or in quarantine. *Science of the Total Environment, 743*, 140803.

Dieng, C., Mberu, B., Dimbuene, Z. T., Faye, C., Amugsi, D., & Aboderin, I. (2022). Biomedical waste management in Dakar, Senegal: Legal framework, health and environment issues; policy and program options. *Cities & Health, 6*(1), 208–222.

Egbenyah, F., Udofia, E. A., Ayivor, J., Osei, M. M., Tetteh, J., Tetteh-Quarcoo, P. B., & Sampane-Donkor, E. (2021). Disposal habits and microbial load of solid medical waste in sub-district healthcare facilities and households in Yilo-Krobo municipality, Ghana. *PLoS One, 16*(12), e0261211.

Farrokhi-Asl, H., Makui, A., Jabbarzadeh, A., & Barzinpour, F. (2020). Solving a multi-objective sustainable waste collection problem considering a new collection network. *Operational Research, 20*, 1977–2015.

Geyer, H. S., Jr., & Van Lille, G. (2024). The medicalisation of solid waste: coordination challenges of domestic biomedical risk waste in illegal dumping sites in informal settlements in Cape Town. *Habitat International, 146*, 103042.

Giglio, E. M., Ryngelblum, A., & de Sousa Jabbour, A. B. L. (2020). Relational governance in recycling cooperatives: A proposal for managing tensions in sustainability. *Journal of Cleaner Production, 260*, 121036.

Hassan, A. A., Tudor, T., & Vaccari, M. (2018). Healthcare waste management: A case study from Sudan. *Environments, 5*(8), 89–101.

Haywood, L. K., Kapwata, T., Oelofse, S., Breetzke, G., & Wright, C. Y. (2021). Waste disposal practices in low-income settlements of South Africa. *International Journal of Environmental Research and Public Health, 18*(15), 8176.

Kain, J. H., Nyakinya, B., Odhiambo, N., Oloko, M., Omolo, J., Otieno, S., Zapata, P., & Zapata Campos, M. J. (2016). Translating policies into informal settlements' critical services: Reframing, anchoring and muddling through. *Public Administration and Development, 36*(5), 330–346.

Kashyap, S., Ramaprasad, A., & Sastry, N. B. (2020). Waste quarantine to reduce COVID-19 infection spread. *The International Journal of Health Planning and Management, 35*(5), 1277.

Laouar, M., Hamad, Z., & Eom, S. (2019). *Towards blockchain-based urban planning: Application for waste collection management*. Cairo, Egypt: *On Information Systems And Technologies*.

Middha, B., & Horne, R. (2024). Countering exclusionary infrastructure in apartment waste management: Towards a relational place-based governance in Victoria. *Urban Sustainability, 4*(1), 9.

Nanda, S., & Berruti, F. (2021). Municipal solid waste management and landfilling technologies: A review. *Environmental Chemistry Letters, 19*(2), 1433–1456.

Ochieng, W. O. (2019). Solid waste management in the informal settlements: A land use planning perspective. *World Environment, 9*(1), 19–27.

Opoko, A. P., & Oluwatayo, A. A. (2016). Private sector participation in domestic waste management in informal settlements in Lagos, Nigeria. *Waste Management & Research, 34*(12), 1217–1223.

Rahman, M. M., Bodrud-Doza, M., Griffiths, M. D., & Mamun, M. A. (2020). Biomedical waste amid COVID-19: Perspectives from Bangladesh. *The Lancet. Global Health, 8*(10), e1262.

Rodseth, C., Notten, P., & Von Blottnitz, H. (2020). A revised approach for estimating informally disposed domestic waste in rural versus urban South Africa and implications for waste management. *South African Journal of Science, 116*(1-2), 1–6.

Shammi, M., Behal, A., & Tareq, S. M. (2021). The escalating biomedical waste management to control the environmental transmission of the COVID-19 pandemic: A perspective from two South Asian countries. *Environmental Science & Technology, 55*(7), 4087–4093.

Singhdeo, A. K., Tripathy, S., & Singhal, D. (2024). Scenarios of waste management in the Indian context: A brief review on E-waste and biomedical waste. *Journal of Physics: Conference Series, 28*(1), 012039.

Squire, J. N. (2013). Biomedical pollutants in the urban environment and implications for public health: A case study. *International Scholarly Research Notices, 2013*(1), 497490.

StatsSA. (2020). *Statistics South Africa general household survey*. StatsSA.

Yadav, I., Juneja, S. K., & Chauhan, S. (2015). Temple waste utilization and management: A review. *International Journal of Engineering Technology Science and Research, 2*, 14–19.

Zikhathile, T., & Atagana, H. (2018). Health care risk waste generated in domestic settings through home health: A review of the health and environmental impact in South Africa. In *Proceedings of the 10th Int'l Conference on Advances in Science, Engineering, Technology & Healthcare (ASETH-18)* (pp. 57–65). Cape Town, South Africa: Eares Conferences.

Glossary

Autoclaving: A steam-based sterilization process used to disinfect infectious waste, reusable instruments, and selected personal protective equipment (PPE) before disposal or reuse.

Biohazard: A biological agent or contaminated material that poses a risk to human health and requires special handling and containment.

Biomedical waste: Waste generated during diagnosis, treatment, immunization, or related research activities involving humans or animals, including contaminated materials, sharps, and microbiological waste.

Bioremediation: The use of microorganisms or their enzymes to degrade pollutants and reduce contamination in the environment.

Capacity building: The process of developing human, technical, and institutional capabilities to improve waste management systems.

Chemical disinfection: The application of chemical agents to inactivate pathogens in waste or on contaminated surfaces.

Circular economy: An economic model that promotes reuse, recovery, recycling, and reduced dependence on virgin resources.

Collection: The systematic removal of segregated waste from generation points for transfer to storage, treatment, or disposal units.

Compliance: Adherence to legal, regulatory, and institutional requirements governing waste handling, treatment, transport, and disposal.

Contaminated PPE: Personal protective equipment such as masks, gloves, gowns, and face shields that have been used in potentially infectious settings and require controlled disposal.

COVID-19 waste: Waste generated from COVID-19 testing, treatment, isolation, and infection-prevention activities, including used PPE, swabs, and contaminated consumables.

Decontamination: The process of removing or inactivating infectious agents to make waste, equipment, or surfaces safe for handling.

Disposal: The final elimination of treated waste by approved methods such as landfill, deep burial, or controlled management of incineration residues.

Environmental impact assessment: A structured evaluation of the potential environmental and public health effects of waste treatment and disposal practices.

E-waste from healthcare: Discarded electronic equipment generated in healthcare settings that requires separate management, particularly when contaminated or obsolete.

Healthcare facility: Any institution that provides diagnostic, therapeutic, preventive, or laboratory services and generates healthcare waste.

Healthcare waste management system: The complete operational framework for waste segregation, collection, storage, transport, treatment, and disposal.

Hazardous waste: Waste that may be infectious, toxic, chemical, radioactive, or otherwise capable of causing harm to human health or the environment.

Healthcare waste: Waste produced by healthcare facilities, laboratories, research institutions, and related services; the majority is nonhazardous when properly segregated.

Incineration: A high-temperature combustion process used to destroy waste and pathogens, with ash and emissions requiring controlled management.

Infection control: Measures implemented to prevent the transmission of infectious agents in healthcare and waste management settings.

Infectious waste: Waste containing pathogens in sufficient quantity to cause disease in humans or animals.

Isolation waste: Waste generated from the care of patients isolated because of suspected or confirmed infection.

Microbiological waste: Laboratory cultures, stocks, specimens, and other materials contaminated with microorganisms, including related vaccine waste.

Microwaving: A nonincineration treatment method that uses microwave energy to heat and disinfect waste.

Nonhazardous waste: General waste generated in healthcare facilities that does not present additional risk when segregated and managed appropriately.

On-site storage: Temporary storage of segregated waste within a healthcare facility before treatment or transfer.

Pathological waste: Human or animal tissues, organs, body parts, and fluids generated during diagnosis, treatment, or research.

Personal protective equipment (PPE): Protective clothing and devices worn to reduce exposure to infectious agents, chemicals, or other hazards.

Pollutants: Harmful substances released into the environment as a result of improper waste treatment or disposal.

Public health protection: The prevention of disease transmission, injury, and environmental contamination through safe waste management practices.

Quarantine waste: Waste produced in quarantine settings or households under quarantine due to potential exposure to infection.

Regulated medical waste: Waste requiring special handling because it may be infectious or physically hazardous.

Resource recovery: The recovery of usable materials, energy, or products from waste streams through approved technologies.

Safe handling: Procedures designed to prevent exposure, spills, injuries, and contamination during waste handling.

Segregation: The separation of waste at the point of generation into designated containers according to type and hazard level.

Sharps: Objects capable of puncturing or cutting skin, such as needles, blades, scalpels, and contaminated broken glass.

Source segregation: The separation of waste immediately at the point of generation, considered the most critical step in biomedical waste management.

Standard operating procedure (SOP): A documented set of instructions describing how a task should be performed consistently and safely.

Storage: The temporary retention of waste before treatment, transport, or final disposal.

Sustainable waste processing technologies: Treatment approaches designed to reduce environmental impact while improving efficiency, safety, and resource recovery.

Treatment: Any physical, chemical, or biological process that reduces the infectious or hazardous nature of waste before disposal.

Transport: The movement of waste within or outside a facility under conditions that prevent leakage, exposure, and contamination.

Waste minimization: Measures aimed at reducing the quantity and hazardous nature of waste through prevention, reuse, and efficient resource management.

Waste valorization: The conversion of waste into valuable products such as energy, secondary raw materials, or other useful outputs.

Waste-to-energy: The conversion of waste into usable energy through thermal or biological processes.

Index

Page numbers followed by "*f*" and "*t*" indicate, figures and tables respectively.

9780443218415